Encoded

The Secret Language Of Nature

GEW Reports & Analyses Team

Global East-West (London)

Hichem Karoui (Editor)

Global East-West

Contents

Chapter 1

Deciphering Earth: An Introduction

Setting the Stage: Why Earth's Language Matters

Understanding Earth's natural processes is crucial for ensuring environmental protection and long-term sustainability. The intricate web of interconnected systems on our planet, including the atmosphere, hydrosphere, lithosphere, and biosphere, all play a critical role in supporting life as we know it. By deciphering the language of Earth, we gain valuable insights into how these systems operate and interact with each other. This understanding is fundamental for making informed decisions that will safeguard the environment for future generations. When we grasp Earth's language, we can better comprehend the impacts of human activities on natural ecosystems, climate patterns, and geological processes. Armed with this knowledge, we can develop strategies to mitigate environmental degradation, reduce carbon emissions, preserve biodiversity, and sustainably manage natural resources. Furthermore, unlocking the secrets of Earth's language provides opportunities to advance technologies that harness renewable energy, protect endangered species, and promote ecological balance. By delving into this linguistic framework, scientists, policymakers, and communities can work together to address environmental challenges and pave the way for a more sustainable future. In essence, understanding Earth's language is not just an academic pursuit; it is a pivotal step towards creating a harmonious relationship between humanity and the natural world.

Overview of Earth's Complex Systems

The study of Earth's complex systems encompasses a diverse array of interconnected components that shape the planet's behavior and evolution. From the intricate web of atmospheric layers to the dynamic energy exchanges within the lithosphere, understanding Earth's complex systems requires a multidisciplinary approach. At the heart of this complexity lies the intricate interplay between Earth's spheres - the geosphere, hydrosphere, atmosphere, biosphere, and anthroposphere, each influencing and being influenced by the others. The geosphere comprises the solid Earth, from the crust to the core, driving geological processes such as tectonics, volcanism, and erosion. Meanwhile, the hydrosphere encompasses all forms of water on Earth, including oceans, rivers, lakes, and groundwater, shaping the landscape and regulating global climate dynamics. The atmosphere, composed of various gases and suspended particles, governs weather and climate patterns, contributing to Earth's dynamic and ever-changing meteorological phenomena. In concert with these spheres, the biosphere represents the realm of life, where organisms interact with their environment, shaping ecosystems and playing a crucial role in global biogeochemical cycles. Finally, the anthroposphere pertains to human societies and their impact on the natural world, underlining the significance of human-environment interactions in shaping Earth's future. Understanding the complexities of these interwoven systems is essential for elucidating Earth's language and how it communicates its history, present state, and potential future outcomes. It requires a holistic perspective that considers the intricate feedback loops, nonlinear behaviors, and emergent properties inherent in Earth's natural processes. By delving into the interconnectedness of Earth's spheres and systems, researchers gain valuable insights into the planet's functioning and the impacts of human activities. This comprehensive understanding is fundamental to interpreting the signals and messages encoded within Earth and harnessing this knowledge for sustainable coexistence and informed decision-making.

Historical Perspectives on Earth Studies

Throughout history, the study of Earth has been an essential pursuit for humanity. From ancient civilizations to modern scientific endeavors, understanding the language of the Earth has been a continual quest. The roots of Earth studies can be traced back to ancient cultures who sought to interpret the elements and natural phenomena as a means of survival and spiritual connection. These early civilizations observed the movements of celestial bodies, tracked weather patterns, and interpreted geological formations, laying the groundwork for our present understanding of Earth's complex systems. As human knowledge advanced, so did our methods of observing and interpreting the Earth. During the Renaissance, there was a remarkable shift towards empirical observation and systematic inquiry, leading to the birth of modern Earth sciences. Pioneers like Galileo Galilei, Nicolaus Copernicus, and Johannes Kepler revolutionized our understanding of Earth's place in the cosmos, laying the foundation for future advancements in Earth studies. In more recent times, the interdisciplinary nature of Earth studies has gained prominence, recognizing that the language of the Earth cannot be fully understood through a single lens. Integrating geology, meteorology, biology, chemistry, and physics, modern Earth studies have unveiled the intricate web of interconnected systems that govern our planet. Additionally, the advent of new technologies, such as satellite imaging, remote sensing, and computational modeling, has expanded our ability to observe and interpret Earth's dynamic processes with unprecedented accuracy and detail. Furthermore, historical events, such as the Industrial Revolution and the impacts of human activity on the environment, have propelled a new urgency in understanding the language of the Earth and its fragility. The historical perspectives on Earth studies serve as a reminder of our ongoing responsibility to preserve and cherish the intricate and delicate balance of Earth's ecosystems. By delving into the historical context of Earth studies, we gain a profound appreciation for the evolution of our understanding and an acute awareness of the pressing need to decipher the language of the Earth for the sustainable future of our planet.

The Interdisciplinary Nature of Decoding Earth

Deciphering the language of Earth is not a task that falls solely within the domain of one scientific discipline. Rather, it demands a collaborative and interdisciplinary approach that draws upon the expertise of multiple fields. Geology, meteorology, biology, astronomy, ecology, and anthropology are just a few of the disciplines that converge in the effort to decode the intricate messages embedded in the Earth's systems. By embracing this interdisciplinary nature of Earth deciphering, researchers can gain a more comprehensive understanding of the interconnectedness of the planet's various elements. Geologists contribute their knowledge of rock formations, strata, and minerals to unveil the stories hidden in the Earth's crust. Meteorologists analyze weather patterns and climatic shifts to decode the atmospheric language, providing valuable insights into the changing dynamics of our planet's climate. Biologists and ecologists study the intricate web of life forms and ecosystems, shedding light on the interwoven relationships between species and the environment. Astronomers extend their gaze beyond Earth, exploring celestial bodies and cosmic phenomena to comprehend the broader context within which our planet exists. Anthropologists delve into the historical interactions between human societies and the Earth, offering cultural perspectives on the interpretation of natural phenomena. The collaborative efforts of these diverse fields enable a holistic approach to decoding Earth's language, enriching our understanding of the planet's past, present, and future. Embracing the interdisciplinary nature of Earth deciphering also brings together a wide array of methodologies and techniques. From satellite imaging and geographical information systems to genetic sequencing and isotope analysis, the toolkit for deciphering Earth's language is as diverse as the disciplines involved. Advanced technologies such as LiDAR (Light Detection and Ranging) and GNSS (Global Navigation Satellite Systems) provide detailed insights into Earth's topography and movements. Meanwhile, traditional field-work and hands-on observation remain integral in grounding scientific discoveries in real-world environments. The complex and evolving nature of Earth's language necessitates an adaptive and flexible approach to research and analysis. Through the convergence of interdisciplinary expertise and methodological diversity, the endeavor to decipher Earth's language becomes not only more robust, but also more resilient in the face of the ever-changing complexities of our planet.

Tools and Techniques for Earth Deciphering

As we delve into the fascinating realm of Earth deciphering, it is essential to recognize the pivotal role played by an array of sophisticated tools and techniques that enable us to unravel the complex language of our planet. Throughout history, advancements in technology have empowered scientists and researchers to gain unprecedented insights into the intricate workings of Earth's natural systems. From remote sensing technologies such as LiDAR and satellite imaging to ground-based instrumentation like seismographs and weather stations, the tools at our disposal continue to expand the frontiers of our understanding. Additionally, the advancement of Geographic Information Systems (GIS) has revolutionized the way spatial data is analyzed and interpreted, offering a powerful platform for integrating and visualizing diverse Earth-related information. Furthermore, the emergence of machine learning and artificial intelligence has brought forth new capabilities in modeling and predicting environmental patterns, providing invaluable support in the interpretation of Earth's language. Cutting-edge laboratory techniques, including isotopic analysis and DNA sequencing, have also opened up new avenues for decoding ancient ecological and geological records, shedding light on Earth's history and evolution. The integration of multiple disciplines, such as geology, ecology, climatology, and astronomy, has facilitated a holistic approach to Earth deciphering, underscoring the interconnectedness of natural processes. Utilizing interdisciplinary approaches, researchers can leverage the strengths of each field to construct a comprehensive narrative that embodies Earth's ever-evolving language. Moreover, the utilization of crowdsourced data from citizen science initiatives has broadened the scope of Earth deciphering, allowing for the collection of extensive datasets and fostering public engagement in environmental research. Looking ahead, the ongoing development of advanced sensor networks, autonomous drones, and next-generation analytical instruments promises to elevate the precision and scale of Earth deciphering endeavors, paving the way for groundbreaking discoveries and novel insights into the language of our planet.

Case Studies: Successes in Earth Communication

In the pursuit of deciphering Earth's language, numerous success stories have emerged, showcasing the effectiveness of various approaches and methodologies. One such case study revolves around the monitoring and analysis of seismic activity to predict volcanic eruptions. Through the meticulous examination of ground vibrations and geological shifts, scientists were able to identify subtle patterns that heralded imminent volcanic events, thereby enabling timely evacuation and disaster management measures. This breakthrough not only saved countless lives but also highlighted the potential for Earth's language to offer crucial insights into natural calamities. Another noteworthy success in Earth communication lies in the field of climatology, where advancements in satellite technology and computational modeling have revolutionized weather forecasting and climate trend analysis. By monitoring atmospheric variables and oceanic currents, researchers have made significant strides in understanding and predicting complex weather systems, enhancing our ability to prepare for and mitigate the impact of extreme weather events. Furthermore, the study of animal behaviors has unveiled compelling instances of successful Earth communication. Ethologists have decoded intricate communication signals among species, shedding light on the nuanced ways in which animals convey information about environmental changes, impending hazards, and social dynamics. These case studies emphasize the tangible benefits of unraveling Earth's language, ranging from societal safety and environmental stewardship to wildlife conservation and sustainable resource management. Moreover, they underscore the interdisciplinary nature of Earth communication, where collaborations between geologists, meteorologists, biologists, and technologists yield transformative results. As we delve deeper into these successes, it becomes evident that decoding Earth's language transcends scientific curiosity; it is a paramount endeavor with far-reaching implications for humanity's coexistence with the planet. These triumphs propel us forward, instilling confidence in our ability to harness Earth's language for the collective betterment of society and the environment.

Challenges and Limitations in Interpreting Natural Data

Successfully deciphering Earth's language is a task that comes with its own set of challenges and limitations. While the progress made in understanding natural data has been tremendous, there are various obstacles that researchers and scientists encounter in their efforts to interpret Earth's communication. These challenges not only stem from the inherent complexity of natural systems but also from the limitations of current technological and methodological approaches. One of the primary hurdles in interpreting natural data lies in the sheer interconnectedness of Earth's intricate web of systems. The interactions between various elements such as the atmosphere, oceans, landmasses, and living organisms create a complex tapestry that can be challenging to unravel. Moreover, Earth's systems operate on multiple temporal and spatial scales, making it difficult to capture and analyze data comprehensively. Another significant challenge arises from the unpredictability and variability inherent in natural processes. Environmental phenomena such as weather patterns, ecological dynamics, and geological events often exhibit non-linear behaviors and sudden changes, making it arduous to predict and interpret their implications accurately. Furthermore, limitations in data quality and quantity pose significant challenges to researchers. Acquiring comprehensive and high-fidelity data from remote or inaccessible locations, as well as maintaining long-term datasets, can be logistically and financially demanding. Additionally, the integration and synthesis of multi-source data, including satellite observations, ground-based measurements, and computational models, present technical and computational challenges that must be addressed. Methodological limitations also contribute to the challenges in interpreting natural data. Analytical techniques and modeling approaches may face constraints in accurately representing the complexities of natural systems, leading to uncertainties and biases in interpretations. Addressing these challenges requires a multi-faceted approach that encompasses advancements in sensor technologies, data processing and analysis methods, modeling frameworks, and interdisciplinary collaborations. Overcoming these limitations will shape the future of Earth language research and pave the way for a more comprehensive understanding of our planet's communication.

Future Directions in Earth Language Research

As we navigate the intricate web of challenges and limitations in interpreting natural data, it becomes increasingly clear that the future of Earth language research holds both promise and complexity. The journey toward understanding and deciphering the myriad languages spoken by our planet demands a concerted effort to explore new horizons and embrace innovative approaches. In this quest for knowledge, several key directions emerge as essential pathways for advancing Earth language research. One pivotal direction lies in the development of advanced technology and analytical tools. The continued evolution of remote sensing technologies, artificial intelligence, and machine learning algorithms presents unprecedented opportunities to delve deeper into the intricate messages embedded within Earth's systems. By harnessing cutting-edge instruments and methodologies, scientists can enhance their capacity to decode Earth's language with greater precision and depth than ever before. Simultaneously, interdisciplinary collaboration stands as a cornerstone for the future of Earth language research. Embracing diverse perspectives and expertise from fields such as geology, climatology, biology, astronomy, and anthropology fosters a holistic approach to unraveling the complex narratives woven into our planet's fabric. Breaking down traditional barriers between disciplines and integrating knowledge streams is essential for developing comprehensive understandings of Earth's multifaceted language. Moreover, the exploration of ancient records and indigenous knowledge systems offers a rich avenue for enriching Earth language research. By delving into historical chronicles, oral traditions, and indigenous wisdom, researchers can gain valuable insights into past interactions with Earth's language and tap into reservoirs of untapped knowledge. Harnessing these timeless sources of wisdom can illuminate previously overlooked dimensions of Earth communication, paving the way for a more nuanced comprehension of the planet's dialogue. A paramount aspect of future research involves addressing the global implications of Earth's language. Understanding the interconnectedness of natural systems on a planetary scale necessitates a shift toward global perspectives that transcend national boundaries. Collaborative efforts to monitor and interpret Earth's language patterns on a worldwide scale are imperative for devising strategies to mitigate environmental crises, anticipate natural disasters, and sustainably manage Earth's resources. The road ahead in Earth language research is fraught with challenges, yet brimming with possibility. By embracing technological advancements, cultivating interdisciplinary partnerships, exploring ancient wisdom, and adopting a global outlook, researchers can forge new frontiers in deciphering Earth's enigmatic language, thereby deepening our relationship with the planet and fostering a harmonious coexistence.

The Language of the Elements

As we prepare to delve into the intricate realms of the earth's elemental language, it is vital to recognize the foundational significance of comprehending these fundamental building blocks of our world. The language of the elements encompasses the complex interplay between earth, air, water, and fire, showcasing the harmonious synergy that sustains life on our planet. In this upcoming chapter, we will embark on a captivating exploration of the unique dialect spoken by each element, unveiling the profound impact they have on ecosystems, weather patterns, geological formations, and human civilization. By gaining an in-depth understanding of the language of the elements, we aspire to decipher the underlying messages that shape our natural environment, leveraging this knowledge to fortify our connection with the Earth. Through the systematic analysis of elemental interactions, we aim to embrace a holistic perspective that transcends individual components, unraveling the intricate web of relationships that underpin the Earth's dynamic systems. Additionally, we seek to shed light on the historical significance of elemental communication, tracing its influence across ancient cultures, folklore, and spiritual beliefs. Furthermore, we will examine the modern applications of elemental language research, exploring how advancements in technology and interdisciplinary collaborations continue to expand our comprehension of these primal forces. As we anticipate our deep dive into the language of the elements, we encourage readers to approach this journey with curiosity and a readiness to marvel at the awe-inspiring intricacies of nature's vocabulary. Reflecting on the profound revelations to come, let us take this moment to ponder the profound interconnectedness of all things, igniting our anticipation for the enlightening exploration of the elemental realm.

Conclusion: The Implications of Understanding Earth's Language

Understanding the language of Earth is a pivotal endeavor that carries profound implications for humanity. By unraveling the subtle messages embedded in nature, we gain insights that can reshape how we interact with our environment and guide our stewardship of the planet. As we conclude our exploration into this intricate subject, it becomes apparent that the implications of understanding Earth's language extend far beyond the realms of scientific curiosity. One of the foremost implications lies in the realm of environmental conservation and sustainability. By comprehending the language of the elements, ecosystems, and geophysical processes, we are better equipped to mitigate the impact of human activities on natural systems. This knowledge empowers us to develop measures that promote biodiversity, safeguard vulnerable habitats, and address climate change with greater precision and efficacy. Furthermore, a nuanced understanding of Earth's language fosters an ethos of responsible environmental management, inspiring innovative strategies for sustainable resource utilization and waste reduction. Moreover, the ramifications of deciphering Earth's language resonate across various spheres of human society. Insights gleaned from the study of Earth's complex systems can inform urban planning, infrastructure development, and disaster preparedness. By integrating our understanding of Earth's language into these domains, we can build resilient communities that harmonize with the natural landscape and are less susceptible to environmental upheavals. Furthermore, the applications of this knowledge extend to agriculture, where it can facilitate practices that optimize yields while preserving soil health and minimizing adverse impacts on neighboring ecosystems. Beyond practical applications, the implications of understanding Earth's language also have the potential to inspire a profound shift in human consciousness. By delving into the intricate dialogues of the natural world, we cultivate a deeper appreciation for the interconnectedness of all life forms and the exquisite balance that sustains the biosphere. This awareness can spur a reevaluation of societal values, prompting a reconnection with the natural world and fostering a sense of responsibility towards future generations. In conclusion, the journey to decipher Earth's language leads us to recognize its significance in shaping the trajectory of human civilization. As we navigate the complexities of an ever-changing world, the wisdom derived from understanding Earth's language serves as a guiding beacon, illuminating pathways towards a more harmonious coexistence with our planet.

Chapter 2

The Language of the Elements

Elemental Language

The concept of elements extends far beyond the realm of chemistry and physics; it encompasses a rich and intricate language that has shaped the very fabric of our world. Elements can be seen as the fundamental vocabulary with which the universe articulates itself. Much like words combine to form sentences, atoms and molecules come together to create compounds and substances, each conveying its own unique message. The aim of this chapter is to delve into the intricacies of this elemental language, exploring how the arrangement of protons, neutrons, and electrons - along with their interactions - creates a diverse lexicon that underpins the natural world. By understanding the basic vocabulary of elements, we lay the foundation for deciphering the complex narratives written across the landscape of science and nature.

Atoms and Molecules: The Basic Vocabulary

The world of atoms and molecules forms the very foundation of our understanding of chemistry and the behavior of matter. Atoms, the building blocks of all elements, are composed of a nucleus containing protons and neutrons, surrounded by electrons that orbit in specific energy levels. These tiny particles come together to form molecules, which are the essential units of chemical compounds. Within these molecules, atoms bond with one another through sharing or exchanging electrons, creating a vast array of substances with unique properties and behaviors. Understanding the structure and behavior of atoms and molecules is crucial to comprehending the interactions and transformations that occur in the natural world.

Subatomic particles, such as quarks and leptons, are the fundamental components of protons, neutrons, and electrons. Each element is defined by the number of protons in its nucleus, known as the atomic number. Isotopes of an element have the same number of protons but differ in their numbers of neutrons, leading to variations in atomic mass. This diversity allows for the formation of an incredible diversity of elements, each with distinct characteristics and roles in the composition of matter.

Molecules, on the other hand, represent combinations of atoms held together by chemical bonds. Covalent bonds involve the sharing of electron pairs between atoms, resulting in molecules with stable structures. In contrast, ionic bonds entail the transfer of electrons from one atom to another, leading to the formation of positively and negatively charged ions that attract each other. Hydrogen bonds, though weaker than covalent or ionic bonds, play a crucial role in maintaining the structures of complex molecules such as DNA and proteins.

Moreover, the study of atoms and molecules extends beyond their physical and chemical attributes; it also encompasses their roles in biological processes. The molecular basis of life, including the composition and interactions of biomolecules like proteins, nucleic acids, lipids, and carbohydrates, provides insight into the intricate pathways of cellular function and metabolism. The dynamic nature of molecular interactions underpins the complexities of biological systems, influencing everything from genetic inheritance to the development and functioning of organisms.

In the pursuit of understanding atoms and molecules, researchers utilize various tools and methodologies, ranging from spectroscopy and microscopy to computational chemistry and quantum mechanics. These approaches enable scientists to delve deeper into the microscopic realm, revealing the intricacies of atomic and molecular structures and behaviors. Additionally, advancements in fields such as nanotechnology and materials science rely on a profound grasp of atomic and molecular principles to drive innovations in diverse applications.

This foundational knowledge of atoms and molecules serves as a gateway to exploring the broader realms of chemistry, physics, biology, and countless interdisciplinary fields. As we continue our journey through the language of the elements, the comprehension of this basic vocabulary will pave the way for unraveling the profound mysteries and potentials encapsulated within the elements and compounds that shape our world.

The Periodic Table: An Alphabet of Chemistry

The Periodic Table is a fundamental tool in the study of chemistry, serving as a visual representation of the elements that make up all known matter. Its structure provides a framework for organizing and understanding the properties and behaviors of these elements. In essence, it serves as an alphabet of chemistry, enabling scientists to decode and comprehend the language of matter. The concept of the periodic table was first proposed by Dmitri Mendeleev in 1869, who arranged the elements according to their atomic mass and chemical properties. This arrangement allowed for the identification of patterns and trends among the elements, leading to the prediction of yet-to-be-discovered elements. The modern periodic table organizes elements by increasing atomic number, which reflects the number of protons in an atom's nucleus. Each element is represented by a symbol, with additional information provided about its atomic number, atomic mass, and chemical properties. The periods and groups within the table highlight recurring patterns in elemental behavior, such as the transition from metals to nonmetals and the variation in electronegativity. The layout of the periodic table also showcases the relationships between elements, emphasizing similarities and differences in their characteristics. Moreover, the periodic table serves as a crucial reference for understanding and predicting chemical reactions, as it illustrates the likelihood of elements forming bonds and undergoing transformations. By delving into the periodic table, scientists gain valuable insights into the fundamental building blocks of the universe and can unlock the secrets of matter's composition and behavior. As research and discoveries continue to expand our knowledge of the elements, the periodic table remains an indispensable tool for unraveling the complexities of chemistry and expanding our understanding of the natural world.

Chemical Bonds: Syntax of Stability

The concept of chemical bonds is fundamental to understanding the behavior and stability of matter. At its core, a chemical bond represents the attractive force that holds atoms together in compounds. This force arises from the interaction between the electrons of the atoms involved. Understanding the syntax of stability within chemical bonds involves delving into the different types of bonds and their unique characteristics.

Covalent bonds, for instance, are formed through the sharing of electrons between atoms. This sharing creates a stable configuration for both atoms, leading to the formation of molecules with distinct properties. Alternatively, ionic bonds involve the transfer of electrons from one atom to another, resulting in the formation of positively and negatively charged ions that then attract each other due to electrostatic forces. Furthermore, metallic bonds, prevalent in metals, result from the delocalization of electrons throughout the material, imparting unique properties such as electrical conductivity and malleability.

The syntax of stability within chemical bonds also extends to the concept of bond energy and bond length. Bond energy refers to the amount of energy required to break a bond between two atoms, while bond length denotes the distance between the nuclei of the bonded atoms. Understanding these properties is pivotal in predicting the reactivity and physical properties of substances. Moreover, the study of chemical bonds has provided invaluable insights into the nature of matter, paving the way for the development of numerous applications across various industries.

Advancements in technology have enabled scientists to visualize and manipulate chemical bonds at increasingly smaller scales. Techniques such as scanning probe microscopy and X-ray crystallography have allowed for the direct observation of molecular structures, shedding light on the intricacies of bonding interactions. Additionally, computational methods have played a crucial role in elucidating the behavior of complex molecules and materials, thereby expanding our understanding of chemical bonding.

As we continue to unravel the syntax of stability within chemical bonds, it becomes evident that this realm of study holds immense potential for addressing contemporary challenges. From designing novel materials with tailored properties to elucidating biochemical processes within living organisms, the comprehension of chemical bonds forms the bedrock of modern scientific and technological endeavors. By navigating this intricate linguistic framework of the elemental world, we gain the ability to harness the power of atoms and molecules with unprecedented precision and insight.

Reactions as Dialogues

Chemical reactions can be likened to dialogues between various elements and compounds, each with its own unique language and purpose. Just as individuals communicate to exchange ideas or express emotions, atoms and molecules engage in reactions to form new substances, release energy, or rearrange their structures. In these molecular dialogues, the reactants introduce themselves by bringing their unique properties and energy levels into play.

The process of reaction initiation is akin to the opening lines of a conversation. Reactants come together, often facilitated by a catalyst or external energy input, establishing initial contact and setting the stage for interaction. As the dialogue progresses, bonds are broken and formed, leading to a rearrangement of atoms and the creation of new compounds.

The manner in which these elements combine is akin to a grammatical structure, with specific rules dictating how the reaction proceeds.

Just as the tone and intensity of human conversations vary, chemical reactions can also display diverse qualities. Some reactions occur explosively, reminiscent of heated debates or impassioned arguments, while others unfold gradually, resembling a thoughtful exchange of ideas. The kinetics of a reaction, including the speed at which it occurs and the factors influencing this pace, determine the overall vibe of the 'dialogue' taking place.

Moreover, the products of a chemical reaction serve as the conclusion of this molecular discourse. Each compound formed is like a statement, encoding information about the reactants involved and the conditions under which the reaction took place. These products can go on to participate in further reactions, prolonging the chemical conversation and propagating the exchange of matter and energy.

Understanding reactions as dialogues fosters an appreciation for the intricate and dynamic nature of chemical processes. By recognizing the parallels between molecular interactions and human communication, we gain a new perspective on the fundamental language of chemistry, unveiling the profound significance of reactions as the expressive language of the elements.

Elemental Influence on Earth's Geology

The geological landscape of the Earth is a testament to the enduring dialogue between elements and the planet. From the majestic granite peaks to the intricate network of underground caves, the influence of elements on Earth's geology is profound and far-reaching. At the core of this influence lies the interplay of various minerals, each with its unique elemental composition and properties. The geological processes shaping the Earth's surface are largely dictated by the presence and interaction of these elements. Silicon, for instance, is a dominant player in the formation of rocks such as quartz, feldspar, and mica. Its ability to bond with oxygen atoms to form silicate minerals defines the very fabric of the Earth's crust and contributes significantly to its structural stability. Additionally, the presence of carbon plays a pivotal role in the formation of key geological features. Carbonate rocks, including limestone and marble, are a result of ancient marine life sequestering carbon from the atmosphere and depositing it on the ocean floor. Over time, these deposits undergo geologic processes, giving rise to stunning formations and landscapes. Moreover, the elemental composition of igneous rocks provides crucial insights into the Earth's volcanic history. The presence of trace elements like magnesium, iron, and aluminum in volcanic rocks not only offers clues about the source of magma but also sheds light on the underlying tectonic processes. Furthermore, elements such as sulfur and phosphorus contribute to the diversity of minerals found in hydrothermal vents and associated geological formations. The dialogue between elements and Earth's geology extends beyond just rock formations. Elements play a pivotal role in the creation and alteration of soil profiles, influencing the fertility and composition of terrestrial ecosystems. The presence of vital nutrients such as nitrogen, potassium, and phosphorus shapes the fertility of soil and supports the growth of diverse plant life. Conversely, the accumulation of toxic metals like lead, cadmium, and mercury can pose significant environmental challenges, impacting both the flora and fauna. Understanding the elemental influences on Earth's geology is essential for deciphering the dynamic processes that have shaped the planet over millennia. By unraveling the intricate interactions between elements and the geological landscape, we gain valuable insights into the Earth's history, evolution, and ongoing environmental dynamics.

Biochemical Languages: Elements in Life Forms

Understanding the biochemical languages that life forms utilize to communicate and function is a fascinating exploration into the intricacies of elemental interactions. At the core of these languages are the fundamental building blocks of life – the elements. From the carbon forming the backbone of organic compounds to the oxygen vital for respiration, elements play a crucial role in shaping the biochemical languages of life forms. The precise arrangement and combination of elements within molecules determine the behaviors and functions of life at the molecular level. Hydrogen bonding, covalent bonding, and other elemental interactions are the grammatical constructs that underpin the language of life. These interactions facilitate the creation of complex macromolecules such as proteins, nucleic acids, and carbohydrates. The coding of genetic information and the transmission of neurological signals are also reliant on the interplay of elements within the intricate machinery of living organisms. Furthermore, the essential roles of trace elements, such as iron, zinc, and copper, in enzymatic reactions highlight the nuanced significance of elements in regulating biological processes. Beyond their individual roles, elements participate in dynamic feedback loops as part of larger biochemical pathways, where they act as messengers, catalysts, and structural components. The study of biochemical languages transcends the biological realm and extends into ecological systems, where elements serve as currency for nutrient cycles and energy transfer within ecosystems. This interconnected web of elemental languages shapes the delicate balance of life on Earth. Moreover, by unraveling the elemental vocabulary of life forms, scientists can glean insights into the evolutionary history of species and the adaptive strategies employed by organisms in response to changing environmental conditions. The quest to decipher the biochemical languages of life forms represents a frontier of interdisciplinary research, intertwining fields such as biochemistry, genetics, ecology, and evolutionary biology. As technological advancements continue to expand our analytical capabilities, the depth and complexity of elemental interactions within life forms are gradually being unveiled, opening new avenues for applications in biotechnology, pharmaceuticals, and environmental conservation.

Cultural and Historical Significance of Elements

Throughout history, elements have held profound cultural and historical significance for various civilizations. The ancient Greeks, for example, believed that the universe was composed of four fundamental elements: earth, water, air, and fire. These elements were thought to represent the essential building blocks of all matter and served as a framework for understanding the natural world. Similarly, the Chinese concept of Wu Xing, or the Five Elements, encompassed wood, fire, earth, metal, and water, and played a central role in traditional Chinese philosophy, medicine, and astrology.

Furthermore, the cultural significance of elements can also be seen in religious and mythological contexts. Many indigenous cultures around the world attribute spiritual importance to certain elements, associating them with deities, creation stories, and rituals. For instance, the element of water has been revered in various cultures as a symbol of purification, rebirth, and life-giving abundance. In Norse mythology, the god Thor's control over lightning and thunder underscores the sacredness of elemental forces in shaping human belief systems and narratives.

Elements have also played pivotal roles in the development of human civilizations. The discovery and application of metals like bronze and iron revolutionized tools, weapons, and architectural achievements, catalyzing societal progress and cultural diffusion. Moreover, elements have often been tied to trade and economic prosperity, shaping global exchange networks and influencing geopolitical dynamics. The historical Silk Road, for instance, facilitated the movement of precious goods such as spices, silks, and rare minerals, showcasing how elements played a central role in shaping international relations and forging cultural connections across continents.

In addition to their material significance, elements have also inspired artistic expression and creativity. Throughout art history, the depiction of natural elements has been a recurring motif in paintings, sculptures, literature, and music compositions. Artists have sought to capture the beauty, power, and transformative qualities of natural elements, infusing their work with symbolic meaning and emotional resonance. The interplay between humanity and the elements has thus provided profound inspiration for cultural and artistic movements, reflecting the enduring impact of elemental symbology across diverse societies.

Overall, the cultural and historical significance of elements is deeply intertwined with human civilization, permeating religious, philosophical, economic, and artistic dimensions. By understanding the rich tapestry of meanings attributed to elements throughout history, we gain valuable insights into the collective consciousness of humanity and the enduring legacy of elemental symbolism.

Modern Applications and Technologies

The study of the language of the elements has led to groundbreaking advancements in modern technology and applications across various industries. The understanding of elemental properties and behaviors has paved the way for innovative developments in fields such as material science, energy production, medicine, and environmental protection.

One of the most significant areas where the knowledge of elements has revolutionized modern applications is in the field of materials engineering. Through the manipulation of elemental compositions and structures, scientists have been able to create novel materials with enhanced properties, such as strength, conductivity, and durability. These advancements have found applications in aerospace engineering, electronics, construction, and many other industries, leading to the development of products and infrastructure that were previously unimaginable.

Moreover, the discoveries in elemental language have played a crucial role in the evolution of energy technologies. From harnessing the power of the sun through photovoltaic cells to optimizing the efficiency of fuel cells and batteries, the understanding of elemental interactions has driven the development of clean and sustainable energy sources. This has significantly contributed to the global effort to mitigate climate change and reduce dependence on non-renewable resources.

In the field of medicine, the knowledge of elemental properties has transformed the diagnosis and treatment of diseases. From the development of advanced imaging agents for medical diagnostics to the utilization of nanoparticles for targeted drug delivery, elemental research has greatly expanded the capabilities of modern healthcare. Furthermore, elemental studies have also facilitated the discovery of new therapeutic compounds and biomaterials, offering promising solutions for improving human health and well-being.

Environmental protection and remediation have also benefitted from the insights gained through the study of elemental language. Innovative technologies for pollution control, wastewater treatment, and natural resource conservation have emerged as a result of understanding the behavior of elements in various environmental systems. These approaches are essential for addressing global environmental challenges and ensuring the sustainability of our planet.

Looking forward, the continued exploration of the language of the elements holds immense potential for further technological and scientific advancements. As researchers delve deeper into the fundamental properties of elements and their interactions, new frontiers in nanotechnology, quantum computing, biotechnology, and beyond are expected to emerge. The applications of elemental knowledge are poised to continue reshaping the world, driving progress and innovation across diverse domains.

Future Directions in Elemental Study

As we stand on the precipice of technological advancement, the future directions in elemental study hold immense promise and potential. The rapid evolution of scientific instruments and methodologies has paved the way for groundbreaking discoveries and a deeper understanding of the elemental language. Looking ahead, one of the key avenues for future research lies in the development of advanced computational models that can simulate the behavior of elements at a level of detail previously unattainable. These simulations could unravel the intricacies of atomic and molecular interactions, shedding light on hitherto unexplored aspects of elemental dynamics. Furthermore, the integration of artificial intelligence and machine learning into elemental studies has the potential to revolutionize the field. By analyzing vast datasets and identifying complex patterns, AI-powered systems can accelerate the pace of discovery and facilitate the design of innovative materials with unprecedented properties. Embracing an interdisciplinary approach will also be crucial in shaping the future of elemental study. Collaborations between chemists, physicists, biologists, and engineers hold the key to unlocking new frontiers in material science, environmental sustainability, and biological applications. The convergence of diverse expertise and perspectives can lead to transformative insights and innovations, propelling elemental research into uncharted territories. Moreover, the advent of nanotechnology presents an exciting avenue for exploring the unique behaviors of elements at the nanoscale. By engineering and manipulating materials at the atomic and molecular levels, researchers can unlock novel functionalities and applications with far-reaching implications across industries. Additionally, the pursuit of sustainable and eco-friendly technologies will drive the exploration of alternative materials and elemental configurations. This imperative shift towards environmentally conscious practices underscores the urgency of delving into the potential of renewable resources and eco-friendly elements. Finally, with space exploration becoming an increasingly tangible reality, the study of extraterrestrial elements promises to unveil unprecedented insights into the cosmic abundance of elements beyond our planet. The analysis of meteorites, planetary composition, and celestial bodies offers a window into the elemental compositions of distant realms, opening doors to unparalleled scientific revelations. Indeed, the future directions in elemental study are multifaceted, holding within them the seeds of transformative discoveries, paradigm shifts, and the redefinition of our relationship with the building blocks of the universe.

Chapter 3

Historical Echoes in Weather Patterns

Climatic History

Climatic history serves as a crucial lens through which we gain insights into the complex interplay between past climates and human civilizations. The study of climatic history involves a multidisciplinary approach that integrates geological, biological, archaeological, and historical data to reconstruct past climate patterns and their impact on Earth's ecosystems and societies.

Understanding the dynamics of climatic change requires the examination of a wide range of proxy data, such as ice cores, tree rings, sediment layers, and historical documents. By analyzing these proxies, researchers can infer past temperature fluctuations, precipitation patterns, atmospheric composition, and extreme weather events.

One of the key themes to be explored in this chapter is the influence of climatic variability on the trajectory of ancient societies and civilizations. By investigating how environmental factors, including temperature shifts, droughts, and floods, shaped human migration patterns, agricultural practices, and societal resilience, we gain a deeper appreciation for the intricate relationship between climate and human history.

Furthermore, this chapter will highlight the relevance of palaeoclimatology in providing valuable context for contemporary climate change discussions. By unraveling the intricacies of past climatic fluctuations, we can discern recurring patterns and potential tipping points, offering invaluable perspectives for understanding and mitigating the impacts of current global climate shifts. Through an exploration of pre-industrial climate variations, we can draw parallels to present-day challenges and unveil critical lessons for sustainable resource management and adaptation strategies.

As we delve into the annals of climatic history, we embark on a captivating journey through epochs and eras, uncovering the narratives of environmental upheavals, natural disasters, and resilient human responses. This section seeks to illuminate the significance of climatic history as a cornerstone of interdisciplinary research, shedding light on the profound connections between past climates and the tapestry of human experiences.

The Role of Meteorology in Historical Contexts

Meteorology, the scientific study of the atmosphere and its phenomena, has played a crucial role in understanding historical contexts. Through the ages, societies have been deeply influenced by weather patterns, making meteorology an essential component in unraveling the tapestry of human history. The impact of meteorology on historical events is profound, shaping the development of civilizations, impacting wars and conflicts, and influencing trade routes and exploration.

One of the key aspects of meteorology in historical contexts is its influence on agriculture. Weather patterns directly affected crop yields, leading to periods of abundance or famine. Ancient civilizations closely observed atmospheric conditions to predict rainfall and droughts, enabling them to plan agricultural activities effectively. The science of meteorology also influenced the establishment of calendars and seasonal celebrations, reflecting the intimate connection between weather forecasting and cultural traditions.

Furthermore, meteorological phenomena such as storms, hurricanes, and extreme temperatures have repeatedly altered the course of history. Severe weather events like volcanic eruptions and tsunamis have impacted the rise and fall of empires, leading to mass migrations and reshaping geopolitical landscapes. Understanding the historical significance of meteorological events allows us to appreciate the interplay between human civilization and the natural world.

An intriguing aspect of meteorology in historical contexts is its influence on art, literature, and mythology. Throughout history, depictions of weather phenomena have been prevalent in artistic expressions, serving as metaphors for human emotions and societal upheavals. Mythologies and folklore from various cultures often incorporate elements of weather deities and legends that personify meteorological forces, showcasing the deep reverence and fear humans have held towards atmospheric events.

In addition to influencing human affairs, meteorology has been intricately woven into the annals of explorations and voyages. Mariners and explorers depended on their understanding of prevailing winds, ocean currents, and seasonal weather patterns to navigate uncharted territories and establish trade networks. Historical expeditions were greatly influenced by meteorological knowledge, with advancements in navigation techniques and shipbuilding being driven by the need to comprehend the nuances of weather systems.

As we delve into the historical dimensions of meteorology, it becomes evident that weather and climate exert far-reaching effects on the human experience, shaping the path of civilizations and leaving indelible imprints on the collective memory of humankind.

Decoding Weather Patterns from Ancient Records

Weather patterns play a crucial role in the narrative of human history, leaving lasting imprints on cultures and civilizations. Deciphering these historical weather patterns from ancient records is a formidable task that requires an interdisciplinary approach drawing from climatology, archaeology, geology, and anthropology. Ancient records such as cave paintings, oral traditions, historical manuscripts, and inscriptions on monuments provide valuable clues about past weather phenomena and their impacts on human societies. By analyzing the chemical composition of ice cores from polar regions and deciphering sediment layers in lakes and oceans, scientists can reconstruct ancient climate data. Furthermore, dendrochronology, the study of tree rings, offers insights into past climatic conditions and extreme weather events. Historical accounts of famines, droughts, floods, and storms shed light on the socio-economic consequences of adverse weather, impacting agriculture, trade routes, and urban settlements. Uncovering and interpreting such records enable researchers to piece together the puzzle of ancient weather patterns, leading to a deeper understanding of its profound influence on human civilization. Moreover, the integration of advanced technologies, such as satellite imagery and computational modeling, has revolutionized the study of historical weather patterns. These tools enable scientists to link palaeoclimate data with present-day observations, fostering a comprehensive understanding of long-term climate variability. Through this holistic approach, we are not only able to decode past weather patterns but also predict future trends, contributing to resilience planning and adaptation strategies in the face of climate change. Decoding weather patterns from ancient records offers a unique perspective on the dynamic interaction between humans and their environment and provides valuable lessons for addressing contemporary climate challenges.

Influence of Climate on Early Civilizations

The study of climate and its influence on early civilizations is a fascinating exploration into the interconnectedness of human history with environmental factors. The success, demise, and adaptation of ancient cultures were fundamentally linked to the climatic conditions they experienced. In regions such as Mesopotamia, the Nile Valley, the Indus Valley, and Mesoamerica, the impact of climate on the development of early societies is evident. The availability of water, the predictability of seasonal patterns, and the presence of fertile land were crucial determinants of a civilization's prosperity.

The emergence of agriculture, often considered the bedrock of civilization, was heavily reliant on suitable climatic conditions for crop cultivation. For instance, the regular flooding of the Nile River in ancient Egypt not only ensured the enrichment of soil but also facilitated the development of a sophisticated irrigation system. In contrast, periods of drought or excessive rainfall could lead to crop failures, food shortages, and social upheaval, as seen in the collapse of the Mayan civilization. The profound impact of climate variability on agrarian societies is a recurring theme throughout history.

Furthermore, the spread of disease and the patterns of human migration were intricately linked to climate. Changes in temperature and precipitation influenced the habitats of disease-carrying organisms, affecting the health and survival of ancient populations. Moreover, shifts in climate patterns often led to the displacement of communities, triggering migrations and conflicts over dwindling resources.

The ability of early civilizations to adapt to changing climatic conditions determined their resilience and longevity. The development of advanced architecture, water management systems, and sustainable agricultural practices reflected the successful response to climatic challenges. Conversely, societies that failed to anticipate or cope with environmental changes faced dire consequences, eventually succumbing to the forces of nature.

By examining the entwined relationship between climate and early civilizations, we gain valuable insights into the adaptive strategies employed by our ancestors. This understanding is particularly relevant in the current era of rapid climate change, illuminating the integral role of environmental factors in shaping the course of human history.

Case Studies: Key Weather Events That Shaped History

Throughout history, weather events have played a pivotal role in shaping the course of civilizations and influencing major historical events. This section will delve into significant case studies that showcase the profound impact of weather occurrences on human societies, providing valuable insights into the dynamic interaction between climatic conditions and historical developments.

One such case study revolves around the Little Ice Age, a period of cooling that extended from the 14th to the mid-19th century. This prolonged climatic anomaly had far-reaching consequences across the globe, leading to widespread crop failures, food shortages, and economic hardships. The severe winters and cooler summers profoundly affected agriculture, trade, and social structures, triggering societal upheavals and shaping geopolitical dynamics. Key historical events such as the European famine of the early 1300s and the French Revolution of 1789 bear testimony to the societal repercussions of adverse climatic conditions.

Another compelling case study examines the Dust Bowl of the 1930s, a devastating environmental catastrophe that engulfed the Great Plains of the United States. Prolonged drought, coupled with unsustainable agricultural practices, led to massive dust storms and soil erosion, causing widespread ecological and economic devastation. The human toll was staggering, as countless families were displaced, livelihoods were shattered, and communities were ravaged by the relentless onslaught of natural forces. The Dust Bowl serves as a poignant reminder of the interconnectedness of human activities with the delicate balance of nature, highlighting the profound repercussions of disregarding environmental considerations.

Furthermore, the catastrophic impact of tropical cyclones on coastal regions presents yet another poignant case study. The Great Hurricane of 1780 stands out as one of the deadliest hurricanes in recorded history, causing unparalleled devastation across the Caribbean. With estimates of fatalities ranging from 20,000 to 27,500, this catastrophic event reshaped the demographic and socioeconomic landscape of the affected areas, precipitating long-lasting repercussions for the indigenous populations and colonial powers alike. Similarly, modern-day storms such as Hurricane Katrina in 2005 underscore the enduring vulnerability of human settlements to the fury of nature and the critical importance of disaster preparedness and resilient infrastructure.

These case studies offer compelling narratives that underscore the profound impact of weather events on historical trajectories, providing valuable lessons for contemporary societies in understanding and mitigating the complex interplay between climate and civilization.

Methodologies in Paleoclimatology

Paleoclimatology, the study of past climates on Earth, is a field that employs various sophisticated methodologies to reconstruct and understand historical climatic conditions. By examining natural archives such as ice cores, tree rings, sediment layers, and fossils, scientists can gather vital information about ancient climates and the factors that influenced them. One of the key methodologies in paleoclimatology is the analysis of ice cores. These cylindrical samples drilled from polar ice sheets or glaciers contain layers that provide a chronological record of past climate changes, atmospheric composition, and environmental conditions. Through detailed examination of these ice cores, researchers can infer temperature variations, trace the history of atmospheric gases, and make connections between climate events and human activities. Another prominent technique utilized in paleoclimatology is dendrochronology, the study of tree rings. Tree rings offer valuable insights into past climates by revealing data on precipitation, temperature, and even occurrences of forest fires. By cross-referencing tree ring patterns from various locations, scientists can create regional and global climate reconstructions. Sediment cores extracted from lakes, oceans, and other bodies of water are also crucial to paleoclimatological analyses. These cores act as time capsules, preserving evidence of past environmental conditions such as temperature, rainfall, and nutrient levels. Additionally, the study of fossilized remains provides significant clues about ancient climates and the flora and fauna that existed during different periods. By combining these diverse methods, paleoclimatologists can develop comprehensive models of past climates, unveiling the complex interactions between atmospheric, oceanic, and terrestrial processes. The significance of these methodologies extends beyond academic curiosity, as they contribute to our understanding of current and future climatic trends. Through the synthesis of paleoclimatological data, we gain critical insights into natural climate variability, the impacts of human-induced climate change, and the resilience of Earth's ecosystems. Moreover, this research aids in refining predictive climate models and informing policy decisions aimed at mitigating the effects of environmental degradation. As the urgency of addressing global climate challenges intensifies, the methodologies of paleoclimatology stand as invaluable tools for illuminating the complexities of Earth's climatic history.

Correlating Archaeological Data with Climatic Changes

The correlation between archaeological data and climatic changes provides a fascinating insight into the intertwined relationship between human civilizations and environmental dynamics. By analyzing artifacts, settlements, and ancient landscapes in conjunction with climatological records, researchers can discern patterns of adaptation, migration, and societal evolution linked to fluctuations in climate. This interdisciplinary approach offers invaluable perspectives on how past societies responded to and coped with environmental challenges. One key aspect of this analysis involves studying changes in settlement patterns and land use in response to shifts in climate. For instance, archaeological evidence suggests that certain civilizations migrated or adapted their agricultural practices in response to changing rainfall patterns or temperature shifts. Furthermore, the examination of paleoclimate data alongside archaeological findings helps in identifying periods of cultural flourishing, societal upheaval, or technological innovation associated with specific climatic conditions. This not only enriches our understanding of the past but also provides valuable lessons for contemporary strategies in mitigating the impacts of climate change on human societies. Moreover, the study of ancient civilizations' reliance on natural resources and the ways in which they managed these resources amid changing climate patterns offers pertinent insights for addressing modern-day sustainability challenges. By synthesizing archaeological and climatological data, scholars can illuminate the intricate web of interactions between humans and their environment throughout history. These investigations not only foster a deeper appreciation for the resilience and ingenuity of past societies but also highlight the urgency of adapting to and mitigating the consequences of present-day climate change.

Predictive Models Based on Historical Weather Data

Predictive modeling based on historical weather data represents a critical aspect of climatological research and natural disaster preparedness. By analyzing and understanding past weather phenomena, scientists can develop models to forecast future climate patterns and extreme events with greater accuracy. These predictive models serve as invaluable tools for various industries, government agencies, and communities that rely on weather forecasts to make informed decisions and mitigate potential risks.

Historical weather data mining is an essential first step in developing these predictive models. By examining long-term climate records, scientists can identify recurring patterns, such as temperature fluctuations, precipitation trends, and storm occurrences. Through meticulous data analysis, researchers can discern the interconnected factors influencing these weather patterns, including oceanic currents, atmospheric pressure systems, and solar radiation variations. This comprehensive understanding of historical weather dynamics forms the foundation of building robust predictive models.

Statistical and computational techniques play a pivotal role in constructing predictive models based on historical weather data. Advanced algorithms and machine learning methodologies enable researchers to uncover complex relationships within the vast amounts of historical climate data. By employing regression analyses, time series modeling, and artificial neural networks, scientists can quantify the probabilistic nature of weather occurrences and simulate potential future scenarios. These cutting-edge predictive models facilitate the projection of climate changes, extreme weather events, and their implications over various temporal scales.

Furthermore, incorporating interdisciplinary knowledge into predictive modeling enhances the accuracy and reliability of forecasts. Collaborations between climatologists, geographers, ecologists, and social scientists contribute diverse perspectives to model construction. For instance, integrating ecological data pertaining to flora and fauna responses to climate variations enriches the predictive capabilities of models, particularly in assessing ecosystem vulnerabilities and biodiversity shifts under changing climatic conditions. Additionally, studying historical societal responses to extreme weather episodes aids in simulating potential social and economic impacts of future climate events.

The significance of predictive models based on historical weather data extends beyond scientific research and permeates into vital sectors such as agriculture, urban planning, and disaster management. Farmers rely on accurate seasonal weather forecasts to optimize crop cultivation strategies and minimize agricultural risks. Urban planners utilize long-range climate projections to design resilient infrastructure capable of withstanding future climate-related challenges. Moreover, emergency management agencies leverage predictive models to formulate proactive measures for mitigating the impacts of severe weather events, thereby safeguarding lives and livelihoods.

In essence, predictive models derived from historical weather data represent a cornerstone of modern climate science, empowering societies to prepare for and adapt to the evolving dynamics of our planet's climate. The ongoing refinement and implementation of these models are imperative in addressing the multifaceted impacts of climate change and fostering sustainable resilience across global communities.

Climatic Shifts and Their Implications on Contemporary Climate

As we delve into the realm of historical climatology and weather patterns, it becomes evident that the shifts in climatic conditions throughout history have profound implications on our contemporary climate. The intricate interplay between past climatic events and their impact on today's climate serves as a critical area of study for researchers and scientists alike. By analyzing the historical data and predictive models based on these findings, we can gain valuable insights into the potential trajectory of our current climate. Climatic shifts, whether gradual or abrupt, possess the capacity to shape the environment, alter ecosystems, and influence human societies in multifaceted ways. By understanding these shifts, we can better equip ourselves to comprehend and respond to the challenges posed by contemporary climate change. The parallels drawn between historical climatic shifts and the ongoing changes in our climate underscore the relevance of this exploration. Furthermore, this section will explore the nuanced relationships between natural climatic fluctuations and anthropogenic influences, shedding light on the complex dynamics at play. Through an examination of case studies and empirical evidence, we aim to elucidate the far-reaching impacts of historical climatic shifts on contemporary climate patterns. Additionally, this discussion will address the pressing need for proactive measures and adaptive strategies to mitigate the adverse effects of modern-day climate change. By fusing insights from historical climatology with contemporary observational data, we can construct a comprehensive framework for comprehending the evolving climate landscape. This holistic approach seeks to amalgamate historical perspectives with cutting-edge scientific advancements, offering a cohesive understanding of the intricate web of factors driving contemporary climatic shifts. As we navigate the terrain of climatic shifts and their implications, this chapter endeavors to underscore the urgency of concerted global efforts towards sustainable practices, environmental conservation, and climate resilience. Ultimately, by delving into the fabric of historical climatology and its interface with today's climate, we endeavor to pave the path for informed decision-making and stewardship of our planet's delicate ecological balance.

Summary and Future Perspectives in Historical Climatology

Historical climatology has revealed invaluable insights into the complex relationship between past weather patterns and their impact on contemporary climate. As we summarize the research findings and investigations presented throughout this study, it becomes evident that historical climatology serves as a critical foundation for understanding modern climate dynamics. The convergence of evidence from various fields including meteorology, archaeology, paleoclimatology, and historical records has enriched our comprehension of the intricate interplay between environmental factors and human history.

Looking ahead, the future perspectives in historical climatology offer promising avenues for further exploration and discovery. One key aspect involves advancements in technological tools for data collection and analysis. With the integration of advanced computational models and high-resolution climate proxies, researchers are poised to delve deeper into the annals of Earth's climatic history. These endeavors will not only enhance our capacity to reconstruct past climate dynamics with greater accuracy but also enable us to anticipate and prepare for potential climatic shifts in the future.

Furthermore, collaborative interdisciplinary research initiatives are essential for bridging gaps in knowledge and methodology. By fostering partnerships between climatologists, historians, geographers, and other related fields, we can cultivate a more holistic approach to unraveling the complexities of historical climatology. This inclusive approach will engender comprehensive and multifaceted understandings of the long-term patterns and trends in Earth's climate, ultimately shedding light on the implications for contemporary climate variability and change.

In addition to these technical and collaborative advancements, there is a growing imperative to communicate the findings of historical climatology to wider audiences. Education and public outreach programs can play a pivotal role in raising awareness about the historical context of climate change and its relevance today. By cultivating a greater understanding of the far-reaching impacts of past climatic events, we can foster informed decision-making and action towards building resilient systems and societies in the face of present and future environmental challenges.

In conclusion, the synthesis of historical climatology offers a panoramic view of the interconnectedness between past, present, and future climates. Through careful analysis, innovative methodologies, interdisciplinary collaboration, and effective communication, the trajectory of historical climatology holds immense potential for deepening our understanding of Earth's climatic evolution and, in turn, shaping strategies for adapting to the complexities of contemporary and future climate dynamics.

Chapter 4

The Secret Language of Insects

Entomological Communication

Entomological communication, the study of how insects communicate, is a fascinating and evolving field that sheds light on the intricate ways in which these tiny creatures interact with each other and their environment. Insects utilize a variety of sensory modalities to convey information, including sound, chemical signals, visual cues, and behavioral patterns. Understanding these communication mechanisms is crucial for deciphering insect behavior and ecology.

At the core of entomological communication lies the fundamental concept of why insects communicate. This can be attributed to various reasons such as mate attraction, warning signals, resource location, and defense mechanisms. By delving into the motivations behind insect communication, researchers gain insight into the evolutionary forces that have shaped these diverse communication systems over millennia.

Furthermore, the methods through which insects communicate are equally diverse and intriguing. The repertoire of sounds produced by insects, ranging from clicks and buzzes to chirps and trills, serves as an important mode of communication within and between species. Chemical communication, facilitated through the release of pheromones and other signaling molecules, enables insects to coordinate complex behaviors such as mating and foraging.

Visual signals also play a pivotal role in entomological communication, with insects employing a wide array of colors, patterns, and body postures to convey messages to conspecifics and potential predators. Understanding the nuanced language of visual cues allows researchers to unravel the intricacies of insect social structures and ecological interactions.

Beyond this, behavioral rhythms exhibited by insects, driven by circadian cues and seasonal changes, represent yet another layer of entomological communication. These temporal patterns dictate activities such as foraging, reproductive behavior, and diurnal or nocturnal activities, showcasing the adaptive significance of time-bound communication strategies.

In the broader context of ecosystems, studying entomological communication unveils the pivotal role of insects in maintaining biodiversity and ecosystem dynamics. Their interactions with plants, other organisms, and abiotic factors underline the interconnectedness of natural systems, highlighting the far-reaching implications of insect communication beyond individual species.

As we embark on an exploration of entomological communication, it becomes evident that unraveling the intricacies of insect language provides novel insights into ecological processes, evolutionary dynamics, and the very tapestry of life on Earth.

Decoding Insect Sounds: Clicks, Buzzes, and Chirps

In the intricate tapestry of entomological communication, the acoustic language of insects holds a multifaceted significance. From the gentle hum of bees communicating the location of nectar to the rhythmic chirping of crickets synchronizing their activities, the soundscape of the insect world is a rich repository of information waiting to be deciphered. At the forefront of understanding this auditory code lies the necessity to comprehend the significance of these sounds within the wider ecosystem. Each click, buzz, or chirp embodies a purpose - an intention woven into the fabric of nature's symphony. By analyzing the frequency, duration, and patterns of these sounds, researchers can gain valuable insights into the behavior, mating habits, and environmental adaptations of various insect species.

The process of decoding insect sounds begins with the meticulous documentation and analysis of the acoustic emissions produced by different species. Through the use of specialized recording equipment and high-resolution spectrograms, scientists can capture and visualize the intricate nuances of insect-generated sounds. These 'sound portraits' serve as the foundation for identifying distinctive communication patterns and recognizing the underlying messages conveyed within each species' unique sonic repertoire.

Moreover, the study of insect sounds extends beyond simple auditory perception. It delves into the realm of evolutionary acoustics, shedding light on the adaptive strategies that have enabled insects to thrive in diverse ecological niches. The evolution of specific sound-producing structures, such as stridulatory organs in grasshoppers or tymbals in cicadas, reflects millions of years of biological refinement geared towards intra-species communication and ecological interaction.

Additionally, the proliferation of anthropogenic noise pollution poses a formidable challenge to the acoustic landscape of insect communication. As human-made sounds encroach upon natural habitats, the delicate balance of insect soundscapes is disrupted, impacting crucial behavioral and reproductive activities. Understanding these dynamics is pivotal in formulating conservation efforts that mitigate the detrimental effects of noise pollution on insect communities.

The intersection of technology and entomology has propelled the field of bioacoustics, enabling researchers to delve deeper into the nuanced world of insect sounds. By employing cutting-edge audio analysis software and acoustic monitoring devices, scientists can unravel previously inaccessible layers of insect communication, unlocking a treasure trove of ecological insights. This integration of technology and entomology serves as a testament to humanity's ongoing quest to unravel the elaborate language of the natural world.

Chemical Conversations: Pheromones and Signals

Pheromones, the chemical messengers of the insect world, play a vital role in shaping the behavior and interactions of various species. These complex compounds, secreted by insects to communicate with each other, are involved in a wide array of activities such as mate attraction, territory marking, and warning signals. The intricate chemical conversations mediated by pheromones offer a fascinating glimpse into the sophisticated language of insects.

Insect pheromones are highly species-specific, with each compound tailored to elicit a precise response from conspecifics. For example, female moths release a unique blend of pheromones to attract males for mating, ensuring reproductive success. Similarly, ants use trail pheromones to navigate their environment and coordinate collective activities within the colony. The diversity and specificity of these chemical signals underscore the remarkable adaptations honed through millions of years of evolution.

The investigation of insect pheromones has not only provided profound insights into the behavioral ecology of various species but also led to practical applications in pest management and agriculture. By synthesizing and deploying synthetic pheromones, researchers and farmers have successfully manipulated insect behaviors to disrupt mating patterns and curb infestations, offering a sustainable alternative to conventional pesticides. Additionally, the study of pheromones has shed light on the intricate web of interactions within ecosystems, revealing the delicate balance maintained by these chemical communications.

Furthermore, the significance of pheromones extends beyond the realm of entomology, with implications for human health and well-being. Research into insect pheromones has inspired developments in areas such as biomedicine, wherein pheromone-based therapies have shown promise in addressing conditions like anxiety and mood disorders. Understanding the chemical dialogues of insects may potentially unlock new avenues for innovative medical interventions, harnessing the power of nature's signaling molecules.

As we delve deeper into the world of chemical conversations among insects, we unveil a realm of intricacy and sophistication that rivals our own linguistic complexities. The study of pheromones and signals showcases the remarkable adaptability and ingenuity of the insect kingdom, offering a testament to the awe-inspiring intricacies of the natural world.

Visual Signals: Patterns and Colors in Insect Interaction

In the realm of entomology, visual signals play a pivotal role in the complex language of insects. These visual cues are utilized by myriad insect species to communicate a wide array of messages, from attracting mates to warning predators and establishing territorial boundaries. Perhaps one of the most striking examples of visual communication in the insect world is the colorful and intricate patterns displayed on the wings and bodies of butterflies. These vibrant patterns serve as both camouflage and a means of signaling to other members of their species, contributing to their survival and reproduction. Similarly, the dazzling displays of color exhibited by certain species of beetles and mantises not only serve to deter potential predators but also aid in identifying suitable mating partners. The utilization of color in these instances showcases the sophistication and strategic nature of insect communication. Another fascinating aspect of visual communication in insects pertains to the use of patterns, whether on wings, bodies, or even in the movement of the insects themselves. For instance, the mesmerizing dance performed by fireflies at dusk serves as a visually striking means of attracting potential mates. Additionally, the orchestrated movements of certain social insects, such as ants and termites, convey critical information about foraging paths and impending threats within their colonies. Beyond mere aesthetics, these patterns and movements are integral components of the intricate web of communication that governs insect interactions. Moreover, the ability of certain insect species to perceive polarized light and ultraviolet wavelengths provides them with an expanded visual spectrum, allowing for the conveyance of subtle messages imperceptible to human observers. This heightened sensitivity to light and color further underscores the complexity of visual communication within the insect kingdom. As researchers delve deeper into the realm of visual signaling in insects, they continue to uncover an astonishing diversity of methods and meanings behind these captivating displays. The intricate interplay of patterns, colors, and movements in the context of insect interaction represents an area ripe for exploration and discovery, offering profound insights into the evolutionary adaptions and ecological dynamics shaping the natural world.

Behavioral Rhythms: Circadian Cues and Seasonal Changes

Insects exhibit a fascinating array of behavioral rhythms driven by circadian cues and seasonal changes. The internal biological clocks of insects govern various aspects of their daily activities, from foraging and mating to resting and seeking shelter. Their responsiveness to periods of light and darkness, as well as environmental conditions, underscores the intricate synchronization of insect behavior and life cycles with the natural world. Understanding these rhythms is pivotal in deciphering the language of insects and appreciating their vital role in ecosystems.

Circadian rhythms, also known as daily rhythms, are regulated by endogenous biological clocks that align with the 24-hour cycle of the Earth's rotation. Insects' sensitivity to light and temperature fluctuations prompts distinct behavioral changes throughout the day. For instance, diurnal insects such as butterflies and honeybees are most active during daylight hours, optimizing their foraging and pollination activities. Meanwhile, nocturnal insects such as moths and certain beetles thrive in the cover of darkness, utilizing moonlight and starlight for navigation and mating displays. These orchestrated patterns of activity demonstrate how insects harmonize their behaviors with diel variations in their surroundings.

Additionally, seasonal changes exert profound influence on the behavior and physiology of insects. The transition between seasons triggers pivotal events such as migration, hibernation, reproductive cycles, and developmental transitions. Many insect species showcase remarkable adaptations to survive and thrive in different seasons, showcasing dynamic responses to factors like temperature, humidity, food availability, and photoperiod. These seasonal nuances unveil the intricacy of insect communication and evolutionary resilience. Studying these seasonal shifts provides invaluable insights into ecological dynamics and supports informed conservation efforts.

Furthermore, the interconnected nature of insect behavioral rhythms with other organisms and environmental elements underscores their crucial role in sustaining biodiversity. Insect activities shape food webs, facilitate pollination, regulate pest populations, and contribute to nutrient cycling. By unraveling the temporal patterns and ecological significance of their behaviors, researchers can unlock critical knowledge about ecosystem stability and resilience. Notably, understanding these rhythms enhances our ability to predict and mitigate the impacts of environmental changes, including those arising from human activities.

The study of circadian cues and seasonal changes in insect behavior represents a captivating frontier in entomology, offering boundless opportunities to deepen our comprehension of the intricacies of nature's language. Embracing the rich tapestry of insect rhythms enables us to forge harmonious relationships with the natural world and uphold the delicate balance of the planet's ecosystems.

Ecosystem Interactions: The Role of Insects in Biodiversity

Insects play a fundamental role in maintaining the delicate balance of biodiversity within ecosystems. Their interactions with various elements of an ecosystem, including plants, animals, and other insects, are crucial for the sustainability and resilience of natural communities. As primary decomposers, insects contribute to nutrient recycling by breaking down organic matter, which enriches the soil and sustains the growth of vegetation. In addition, certain insect species, such as bees and butterflies, are essential pollinators, facilitating the reproduction of flowering plants and ensuring the production of fruits and seeds. This critical role in the pollination process directly impacts the diversity and abundance of plant species, thereby influencing the entire food web within a given ecosystem. Furthermore, insects function as both prey and predators, participating in complex food chains and webs that regulate population dynamics and prevent the proliferation of any single species. Their presence serves as a check on potential pest outbreaks and contributes to the overall stability of the ecosystem. Moreover, some species of insects form mutualistic relationships with specific plants, wherein they provide vital services like seed dispersal or protection from herbivores, in exchange for nourishment or shelter. These interdependent associations support the coexistence and interconnectivity of various species, contributing to the overall health and robustness of the ecosystem. Understanding the nuanced roles of insects in biodiversity is instrumental in informing conservation strategies and managing ecological systems. By recognizing their significance in ecosystem dynamics, we can develop effective measures to preserve and protect the intricate web of life that depends on the varied contributions of these often underappreciated organisms.

Human-Insect Dynamics: Coexistence and Conflict

Human-insect dynamics encompass a complex interplay of coexistence and conflict that has persisted throughout human history. Insects, as integral components of ecosystems, play crucial roles in pollination, nutrient cycling, and pest control. However, the relentless expansion of human settlements and agricultural activities has exerted significant pressure on insect populations, often leading to conflicts as they encroach upon human habitats and resources. Understanding the multifaceted nature of these dynamics is essential for maintaining ecological balance and sustainable cohabitation.

Coexistence between humans and insects is fundamentally rooted in acknowledging the need for mutual respect and harmonious cohabitation. Cultures across the globe have recognized the significance of insects in various aspects of their livelihoods, from the revered status of bees in ancient civilizations to the symbolic representations of insects in art and literature. Embracing this coexistence involves recognizing the intrinsic value of insects and the ecosystem services they provide, fostering an ethos of conservation and responsible stewardship of natural habitats, and implementing practices that minimize harm to insect populations. Additionally, promoting public education and awareness regarding the importance of insects can help foster a more harmonious relationship with these remarkable creatures.

Conversely, human-insect conflict arises due to diverging interests and competing needs within shared environments. Agricultural landscapes often bear the brunt of this conflict, as insects are viewed as pests that threaten crop yields and food security. The use of chemical pesticides and intensive agricultural practices further exacerbates these tensions, leading to imbalances in ecosystems and detrimental consequences for biodiversity. Mitigating these conflicts requires a comprehensive approach that integrates ecological principles, sustainable farming practices, and innovative pest management strategies that minimize environmental impact. Additionally, engaging in dialogue with local communities and stakeholders to address concerns and develop collaborative solutions is vital for navigating these complex interactions.

Furthermore, urban environments present unique arenas for human-insect dynamics, as burgeoning cities and human infrastructure impinge upon insect habitats and natural corridors. Encounters with stinging insects, infestations in households, and disease vectors amplify these conflicts, prompting efforts to manage and mitigate their impact. Implementing integrated pest management strategies, creating green spaces that support diverse insect populations, and utilizing architectural designs that minimize insect-human interactions are essential for fostering more sustainable urban ecosystems.

As we navigate the intricate web of human-insect dynamics, it is imperative to recognize the interconnectedness of our fates and the pivotal role that insects play in sustaining the fabric of life on Earth. By adopting a holistic and proactive approach grounded in mutual understanding and coexistence, we can strive towards fostering balanced and mutually beneficial relationships with the diverse spectrum of insect species that share our planet.

Research Methodologies: Tools for Studying Insect Language

Studying the intricate language of insects requires a diverse set of research methodologies and tools to decode their complex communications. One of the fundamental tools used in this field is acoustic monitoring, which involves recording and analyzing the sounds emitted by various insect species. This method allows researchers to identify distinct patterns, frequencies, and temporal variations in insect vocalizations, shedding light on their social interactions, mating rituals, and territorial behaviors. Additionally, visual observations play a crucial role in understanding insect communication. By meticulously documenting the visual signals, such as body movements, wing displays, and color patterns, researchers can gain valuable insights into the subtle nuances of insect language. Another integral aspect of studying insect language is the use of chemical analysis techniques. Pheromones, the chemical signals released by insects, are pivotal in mediating their behaviors, including mating, foraging, and defense. Gas chromatography-mass spectrometry (GC-MS) and other advanced analytical methods enable scientists to isolate, identify, and quantify these chemical compounds, unraveling the intricacies of insect communication through olfactory cues. Furthermore, sophisticated imaging technologies, such as electron microscopy and high-speed videography, provide detailed visualization of minute insect structures and behaviors, facilitating an in-depth understanding of their communication modes. It is also essential to employ molecular biological approaches in deciphering insect language. DNA sequencing, gene expression analysis, and genome editing techniques offer unparalleled insights into the genetic basis of insect communication, unraveling the evolutionary origins and ecological significance of their linguistic patterns. Moreover, interdisciplinary collaborations with experts in bioacoustics, entomology, ecology, and behavioral science enhance the holistic exploration of insect language, fostering a comprehensive understanding of their intricate communication systems. As technology continues to advance, innovative tools such as spectroscopy, machine learning algorithms, and remote sensing applications are revolutionizing the study of insect language, paving the way for groundbreaking discoveries in this captivating field.

Case Studies: Ant Colonies and Bee Dances

The study of insect communication has unlocked a world of intricate behaviors and interconnections, revealing intriguing case studies that provide valuable insights into the language of insects. Ant colonies, renowned for their complex social structures, offer a fascinating landscape for understanding the role of chemical and tactile cues in coordinating activities such as foraging, defense, and reproduction. Through meticulous observation and advanced technology, researchers have delved into the detailed mechanisms underlying ant communication, shedding light on how these minuscule creatures coordinate activities to maintain the colony's resilience and survival. The phenomenon of bee dances, as observed in honeybee hives, represents another remarkable case study in insect communication. These intricate movements and patterns serve as a sophisticated form of information transfer, enabling bees to convey precise details about the location and quality of food sources to fellow members of the colony. By decoding these intricate dances, scientists have gained profound insights into how honeybees navigate their environment and collaborate for the collective benefit of the hive. Case studies of ant colonies and bee dances not only highlight the incredible complexity and effectiveness of insect communication but also underscore the critical role of these interactions in shaping ecosystems and agricultural landscapes. Such investigations hold immense potential for inspiring innovative approaches in fields ranging from biodiversity conservation to sustainable agriculture, as we continue to unravel the secrets of the natural world.

Implications for Ecology and Future Research Directions

The study of insect communication has far-reaching implications for the field of ecology and offers new pathways for future research. Understanding the complex language of insects can provide valuable insights into ecological dynamics, species interactions, and the functioning of ecosystems as a whole. By unraveling the intricate web of insect communication, researchers can gain a deeper understanding of how these tiny organisms shape the natural world and influence larger biological processes. One of the key implications of studying insect communication lies in the realm of conservation biology. As we grasp the nuances of how insects communicate within their communities, we can better comprehend the intricacies of their roles in sustaining biodiversity. This comprehension, in turn, allows us to make informed decisions about conservation strategies aimed at preserving vital ecosystems and preventing species loss. Furthermore, exploring the language of insects opens up new opportunities for pest management and agricultural practices. By understanding the signals and cues used by insect groups, scientists and farmers can develop innovative, environmentally friendly methods to control pest populations and minimize crop damage. Leveraging the knowledge of insect communication can lead to sustainable agricultural practices that reduce reliance on chemical pesticides and foster harmonious coexistence between humans and insects. Looking ahead, the future of research in insect communication holds immense promise. Advancements in technology and interdisciplinary collaborations present exciting prospects for delving deeper into the complexities of insect languages. Integrating fields such as acoustics, chemical ecology, behavioral science, and advanced imaging techniques can propel our understanding of insect communication to unprecedented levels. Moreover, emerging research directions could shed light on the potential applications of insect communication in fields beyond ecology. From bio-inspired technology to medical advancements, the study of insect language may offer novel solutions and innovations that extend far beyond the natural world. However, these future research endeavors will require cross-disciplinary efforts, robust funding, and a commitment to ethical and sustainable scientific practices. Nurturing this burgeoning field of study is essential for unlocking the full spectrum of possibilities that lie within the secret language of insects.

Chapter 5

Animal Behaviors: Messages Through Time

Animal Behavioral Linguistics

Animal behavioral linguistics delves into the intricate and captivating world of communication among non-human species. These fascinating interactions have long been a subject of profound scientific inquiry, shedding light on the diverse ways in which animals express themselves and convey information. At the heart of animal behavioral linguistics lies the exploration of the evolutionary significance and ecological adaptations associated with communication systems. By studying these communication patterns, researchers seek to decode the underlying meaning and purpose behind various signals and behaviors exhibited across the animal kingdom.

In this domain, terminology such as ethology, proxemics, and ethogram are employed to better understand the nuances of animal communication. Ethology involves the scientific study of animal behavior within their natural habitats, aiming to comprehend the adaptive, functional, and evolutionary aspects of behavior. Proxemics refers to the spatial relationships between individuals of the same species during interaction, offering insights into social dynamics and territoriality. An ethogram, on the other hand, is a catalog of all the distinct behavior patterns exhibited by a particular species, comprising a valuable resource for researchers to document and analyze animal communication comprehensively.

Furthermore, the study of animal communication encompasses an assortment of modes, including acoustic, visual, olfactory, and tactile signals. Each mode plays a crucial role in facilitating communication within and between species, reflecting their unique ecological niches and social structures. For instance, the intricate courtship dances of birds or the mesmerizing displays of fireflies illustrate the vivid visual signals utilized in mating rituals. Similarly, the resonant calls of primates echoing through the forest canopy highlight the significance of acoustic communication in establishing group cohesion and defining territories.

Animal behavioral linguistics transcends mere observation, venturing into deeper explorations of cultural transmission and social learning within animal communities. It underscores the cognitive complexity inherent in animal communication systems, underscoring the adaptive advantages and selective pressures that have shaped these behaviors over millennia. By scrutinizing these linguistic intricacies, scientists can unravel the rich tapestry of inter-species interactions, paving the way for a more profound understanding of the animal kingdom's vibrant and enigmatic languages.

The Evolutionary Perspective of Communication

Throughout the history of life on Earth, the communication patterns of animals have undergone a continuous process of adaptation and refinement. From the primordial depths of ancient oceans to the bustling ecosystems of today, the evolutionary perspective of animal communication provides a fascinating insight into the intricate dynamics of survival, reproduction, and social organization. At its core, communication serves as a vital tool for species to navigate their environments, establish social hierarchies, and ensure reproductive success.

One of the fundamental drivers of the evolution of animal communication is the concept of natural selection. Over millennia, organisms that were able to convey and interpret meaningful signals gained distinct advantages in terms of finding food, avoiding predators, and attracting mates. These selective pressures have sculpted a diverse array of communication strategies across the animal kingdom, each finely tuned to suit the specific ecological niche of its practitioners.

The origins of animal communication trace back to prehistoric ancestors, where rudimentary forms of signaling and vocalization emerged as primal responses to environmental stimuli. As species diversified and inhabited diverse habitats, these communication systems underwent substantial refinements. The symbiotic relationship between communication and ecological factors led to the development of specialized signals tailored to address the unique challenges of each habitat, ranging from dense rainforests to expansive savannahs.

Moreover, the interplay between communication and social dynamics played a pivotal role in shaping behavioral repertoires. Social structures within animal communities have given rise to elaborate signaling mechanisms, such as visual displays, vocalizations, and olfactory cues, that serve to maintain cohesion, minimize conflicts, and establish alliances. The hierarchical nature of social organizations has driven the evolution of complex communication systems, with dominant individuals often wielding distinctive signals to assert authority and subordinates utilizing submissive behaviors to prevent confrontations.

Additionally, the significance of evolutionary principles in understanding animal communication extends to the establishment of mate selection criteria. Sexual selection has been a driving force behind the elaboration of courtship displays, songs, and dances that function as indicators of genetic fitness and compatibility. The evolutionary imperative to secure successful mating opportunities has prompted the development of intricate and aesthetically captivating communication rituals, showcasing the profound influence of communication in shaping reproductive success.

As we delve deeper into the evolutionary framework of animal communication, it becomes evident that it represents an intricately woven tapestry of adaptation, coevolution, and behavioral plasticity. The relentless interplay between selective pressures, ecological interactions, and social dynamics has forged an exquisite mosaic of communication systems, each painting a vivid portrait of the evolutionary journey undertaken by countless species across the eons.

Decoding the Basic Elements of Animal Signals

In the intricate web of animal communication, deciphering the basic elements of signals is fundamental to understanding the underlying language spoken across various species. Animal signals can be visual, auditory, olfactory, or tactile in nature, and each embodies a complex set of information that conveys vital messages within their communities. Visual signals, such as body postures, colors, and patterns, play a crucial role in conveying social status, reproductive fitness, and defensive strategies. These signals not only serve as means of expression but also shape the dynamics of social interactions and hierarchies. Auditory signals, ranging from intricate bird songs to the rhythmic drumming of gorillas, aptly demonstrate the diversity and richness of acoustic communication in the animal kingdom. Moreover, olfactory cues, often imperceptible to humans, are pivotal in marking territories, identifying individuals, and signaling reproductive readiness. The subtle language of touch among animals, especially primates and social mammals, communicates comfort, reassurance, and bonding. Understanding these basic elements helps in unraveling the intricate tapestry of animal communication, opening doors to understanding their ecological roles and evolutionary adaptations. Through meticulous observation, scientific inquiry, and advanced technology, researchers delve deep into decoding these signals, shedding light on the covert conversations that have shaped biological processes for millennia.

The Translation of Territoriality and Dominance Displays

Territoriality and dominance displays form a crucial component of animal communication, playing a pivotal role in shaping social interactions and hierarchies within various species. From the majestic movements of big cats marking their territory to the intricate dances of birds asserting dominance, these displays are rich in nuance and significance. By examining and interpreting these behaviors, researchers gain valuable insights into the complex language of the animal kingdom. Dominance displays often involve a series of carefully choreographed postures, vocalizations, and gestures, each carrying its own distinct message. A subtle shift in body posture or the modulation of a vocalization can convey volumes about an individual's status and intentions within the group. Understanding and deciphering these displays requires a keen eye for detail and an appreciation for the subtleties of non-verbal communication. The study of territoriality offers a fascinating glimpse into the intricate ways in which animals demarcate and defend their living spaces. Whether through scent-marking, vocal challenges, or physical confrontations, territorial behaviors play a vital role in defining boundaries and minimizing conflict within a given ecosystem. Furthermore, the translation of territorial cues provides critical insights into resource distribution, mating opportunities, and the preservation of genetic diversity within populations. Researchers employ an array of sophisticated tools and techniques, including behavioral observation, technological monitoring, and bioacoustic analysis, to unravel the intricate tapestry of territoriality and dominance displays across species. Such studies not only enrich our understanding of animal behavior but also shed light on the intricate web of relationships that underpin the natural world. As we delve deeper into the realm of animal communication, the translation of territoriality and dominance displays stands as a testament to the intricate beauty and profound complexity of nature's language.

Mating Rituals: Choreographed Messages

Mating rituals are captivating displays of behavioral communication that underscore the complexity and sophistication of animal courtship. These choreographed messages serve as a crucial method for potential mates to assess genetic fitness, compatibility, and overall suitability in the pursuit of successful reproduction. In the avian world, for instance, the vibrant plumage and intricate dances of peacocks serve as a mesmerizing visual language that signals health and vitality to peahens. Similarly, the graceful aerial performances of various bird species not only demonstrate physical prowess but also indicate an ability to provide for offspring—a compelling form of advertisement to potential partners.

In the aquatic realm, marine creatures engage in elaborate courtship rituals, such as synchronized swimming and mesmerizing color displays, to attract potential mates. The pulsating hues of coral reef fish or the intricately woven nests of pufferfish all contribute to the subtle yet persuasive language of courtship. These behaviors are indicative of genetic quality, parental investment, and reproductive capability, which play pivotal roles in the selection of an optimal mate.

Beyond the realms of terrestrial and aquatic ecosystems, the mammalian world is equally rich with diverse courtship displays. From intricately patterned duets of gibbons to the majestic rutting displays of deer, each ritual carries inherent meaning and significance. The precise timing and coordination of these choreographed displays not only convey the suitability of a potential partner but also serve as mechanisms for assessing the commitment and contribution of a mate in the rearing of offspring.

The study of mating rituals offers profound insights into the evolutionary forces that have shaped these intricate languages of courtship over time. By understanding how these messages have evolved and diversified across different species, we gain invaluable perspectives on the underlying biological and ecological pressures that mold these behaviors. Research in this field further illuminates how environmental factors, competition, and sexual selection have collectively sculpted the rich tapestry of courtship behaviors observed in the natural world.

As we delve deeper into the realm of mating rituals, the remarkable diversity and intricacy of these choreographed messages continue to unveil the marvels of animal communication. These captivating displays not only showcase the remarkable adaptability of species but also shed light on the enduring interplay between innate behaviors and environmental influences in the perpetual dance of life.

Warning Calls and Alarms: The Language of Survival

In the intricate web of animal communication, warning calls and alarms play a vital role in signaling potential threats and ensuring the survival of individuals and groups. Through these vocalizations and non-verbal cues, animals transmit crucial information about the presence of predators or other dangers, allowing their peers to react swiftly and seek safety. This chapter delves into the fascinating intricacies of these communication methods observed across various species and ecosystems.

Warning calls often serve as early detection systems, alerting others to imminent peril. Whether it's the piercing distress cries of birds at the sight of a hawk or the low-frequency vibrations produced by elephants to signal danger, each species has developed unique methods to communicate fear and trigger an organized response. Notably, the urgency conveyed in warning calls can vary, with some alerts prompting immediate flight while others prompt readiness for defensive action.

The study of alarm signaling also unveils the complex dynamics within animal communities. Members not only interpret warnings from their own species but may also recognize signals from different species, acknowledging the universal importance of collaborating against shared threats. Furthermore, research has revealed that certain animals exhibit regional dialects or variations in their alarm calls, suggesting a sophisticated level of communication specific to distinct social groups or environments.

Beyond vocal communications, visual alarms such as specific body postures and gestures are prevalent in many species, providing additional layers of messaging that aid in collective defense strategies. This visual language may involve tail movements, raising fur, or even color changes, each conveying distinct meanings and intensities of danger. These visible signals often complement auditory warnings, offering redundant layers of information exchange for increased accuracy and comprehensiveness.

Understanding the intricacies of warning calls and alarms is not just a biological pursuit; it holds implications for human society as well. By observing how animals coordinate responses to threats through communication, we gain insights into the foundations of teamwork, crisis management, and adaptive decision-making. These learnings can inspire innovative approaches in fields such as emergency response protocols, wildlife conservation, and even cybersecurity. Moreover, recognizing the universality of alarm signals reminds us of the interconnectedness of all life forms and the imperative of coexistence and mutual protection in the face of adversity.

Social Structure and Hierarchal Communication

In the intricate tapestry of wildlife, social structure and hierarchal communication permeate through the lives of many species, shaping their interactions and defining the dynamics of their communities. Whether it's a pride of lions on the African savannah, a troop of chimpanzees in the dense rainforests, or a pod of orcas traversing the vast oceans, the establishment of social hierarchies plays a pivotal role in their survival and prosperity. These structures are often characterized by complex networks of relationships, dominance hierarchies, and behavioral patterns that facilitate cooperation and coexistence. Social structure encompasses divisions of labor, mating systems, and territorial rights, all of which contribute to the cohesive functioning of these animal societies. Within these hierarchies, communication serves as the glue that binds individuals together and maintains order within the group. From subtle gestures and vocalizations to elaborate displays and rituals, animals utilize a diverse array of communication mechanisms to convey social status, maintain peace, establish dominance, and foster alliances. Observing these interactions offers valuable insights into the evolutionary strategies that have enabled these species to thrive in diverse habitats for millennia. Understanding the nuances of social structure and hierarchal communication not only unveils the intricacies of animal societies but also sheds light on our own human social systems. By delving into the depths of these interwoven relationships, we gain a profound appreciation for the significance of social bonds and the symbiotic connections that underpin the fabric of life on Earth. Through continued research and observation, we can unravel the mysteries of social organization and hierarchal communication, paving the way for a deeper comprehension of the natural world and our place within it.

Symbiosis and Cooperation: Cross-Species Dialogues

In the intricate tapestry of the natural world, symbiosis and cooperation play a fundamental role in facilitating cross-species dialogues. From mutualistic relationships to commensalism, various forms of symbiosis foster communication and collaboration between different organisms. These interactions not only shape individual species but also have broader implications for entire ecosystems. Symbiotic relationships often involve the exchange of signals and cues, creating a complex web of interdependent communication that transcends species boundaries.

One remarkable example of symbiosis is the mutualistic relationship between certain species of cleaner fish and larger marine animals like sharks and manta rays. The cleaner fish remove parasites and dead skin from the larger animals, while gaining sustenance and protection in return. This intricate dance of cooperation is choreographed through subtle signals and behaviors, representing a form of cross-species communication that is essential for the survival of both parties involved.

Cooperation among different species also manifests in the context of seed dispersal, where plants rely on various animal species to propagate their seeds. This process involves a sophisticated exchange of signals, as plants develop specific mechanisms to attract and engage with different animal species, such as birds, mammals, or insects, to aid in their reproductive cycle. In turn, these animals receive nourishment from the fruits or seeds they consume and contribute to the dispersion of plant genetic material, highlighting the intricate dialogue that occurs across different realms of the natural world.

Furthermore, symbiosis and cooperation are not limited to visible interactions but also extend into the microbial realm. The intricate associations between microorganisms and larger organisms, such as the gut microbiota in animals, demonstrate a level of communication and cooperation that influences not only individual health but also broader ecological dynamics. The exchange of chemical signals and metabolic byproducts between host organisms and their associated microbiota represents a remarkable form of cross-species dialogue that has significant implications for the functioning of ecosystems.

This dynamic interplay of communication and cooperation across species provides researchers with a rich tapestry of phenomena to explore and understand. Studying these interactions not only sheds light on the intricacies of the natural world but also holds potential insights for fields such as ecology, evolution, and conservation. As we delve deeper into the complex web of connections that define symbiotic and cooperative relationships, we unveil the profound interconnectedness of life on Earth and gain a deeper appreciation for the diverse languages spoken across different species.

Case Studies of Unique Behavioral Patterns Across Different Species

Exploring the spectrum of animal behavior across diverse species illuminates the richness and complexity of non-verbal communication in the natural world. From the intricate dance of a bee's waggle to the majestic choreography of a humpback whale pod, each species possesses its own distinct lexicon, finely attuned to their ecological niche and social dynamics. In the African savanna, the cooperative hunting strategies of the lion pride portray a remarkable display of synchronized communication, with subtle cues and signals orchestrating their pursuit of prey. Contrasting this, the solitary and enigmatic nature of the snow leopard underscores a different language, characterized by elusive encounters and solitary sophistication. The fascinating vocalizations and physical displays of birds, such as the courting rituals of the superb lyrebird or the intricate mimicry of the mockingbird, epitomize the varied and sophisticated forms of avian communication. Evaluating cases like the altruistic behavior of the dwarf mongoose society versus the territorial disputes among chimpanzee groups sheds light on the roles of cooperation and conflict within interspecies interactions. Studying the remarkably diverse communication methods deployed by creatures ranging from insects to marine mammals not only expands our understanding of their intricately woven social fabric but also provides insights into the evolutionary underpinnings of behavioral diversity. Furthermore, advancements in technology, such as bioacoustics and high-resolution imaging, have revolutionized our ability to capture, analyze, and comprehend these intricate behavioral patterns across species. Through comprehensive case studies, we can parse the nuances of animal behavioral communication and delve deeper into the complex tapestry of interspecies dialogues, unraveling the depth of meaning encoded in these interactions and paving the way for future discoveries in the captivating realm of animal behavior.

Future Research and Technologies in Animal Behavior Study

The exploration of animal behavior has always been a fascinating subject, but with advancements in technology, the future holds even more promising opportunities for research. As we delve into the realm of animal behavior, emerging technologies are providing us with unprecedented ways to observe, analyze, and interpret the various facets of animal communication and interactions.

One of the most significant advances lies in the field of bioacoustics. With the development of sensitive audio-recording devices and specialized software, researchers can now capture and analyze intricate vocalizations and other acoustic signals emitted by animals. This technology opens doors to studying animal communication patterns in unprecedented detail, offering insights into the subtle nuances and meanings embedded within their vocal expressions.

Furthermore, the integration of GPS tracking and telemetry systems has revolutionized our ability to monitor animal movements and social dynamics. By deploying miniaturized tracking devices on individual members of animal communities, scientists are able to gain a comprehensive understanding of their ranging behavior, migration patterns, and territorial demarcations. This real-time data collection provides invaluable information for deciphering the intricate language of animal interactions and the underlying motivations driving their behaviors.

Advancements in genetic analysis and molecular biology have also become integral components of animal behavior research. The ability to extract and sequence DNA from fecal matter, hair, or saliva samples has enabled scientists to identify individual animals, examine kinship structures, and investigate genetic contributions to behavioral traits. Such insights shed light on the heritability of specific behaviors and the potential adaptability of species in response to environmental changes.

The application of artificial intelligence (AI) and machine learning algorithms presents an exciting frontier in understanding animal behaviors. These cutting-edge tools can process vast quantities of data and recognize complex patterns that may elude human observation. By harnessing AI, researchers can develop predictive models, detect subtle behavioral cues, and unveil hidden correlations within the extensive datasets amassed from field studies and remote monitoring.

Moreover, the burgeoning field of virtual reality (VR) offers novel prospects for simulating and exploring animal environments. VR platforms enable scientists to immerse themselves in simulated ecosystems, gaining firsthand perspectives of the sensory inputs and stimuli encountered by various species. This immersive experience can inform our understanding of how animals perceive their surroundings and interact within their habitats, facilitating a deeper comprehension of their behavioral responses.

As we venture into the future of animal behavior study, these technological advancements hold tremendous promise in unraveling the intricate tapestry of animal communication, social dynamics, and adaptive strategies. By embracing these innovative tools and methodologies, we stand poised to unlock new dimensions of understanding in the wondrous world of animal behaviors.

Chapter 6

The Language of the Stars

Stellar Linguistics

Stellar linguistics encompasses a diverse array of disciplines that collectively seek to understand the language of the stars. At its core, it examines the foundational aspects of how stars communicate through their patterns and behaviors. By delving into this complex framework, we unravel the celestial grammar through which stars convey their stories across the cosmos. The historical evolution of stellar linguistics reveals the pivotal role played by ancient civilizations in decoding astronomical symbols. From the celestial observations of early cultures to the development of sophisticated telescopes and instruments, humanity's quest to decipher the messages embedded in the stars has been an enduring endeavor. The intricate tapestry of stellar communication intersects with cultural interpretations, as myths and legends from around the world have woven rich narratives based on celestial phenomena. Exploring these interpretations provides valuable insights into the human fascination with the night sky and its enduring influence on literature, art, and beliefs. As we embark on this exploration, we are confronted with the unparalleled diversity of stars, each pulsating with its own unique dialect. From the flickering binary systems to the enigmatic pulsars and quasars, the stars offer a compelling repertoire of linguistic expressions. Moreover, the stellar spectra, analyzed through the lens of spectroscopy, act as a lexicon that enables us to discern the chemical compositions and temperatures of stars, thereby shedding light on their intrinsic messages. The study of interstellar medium further enriches our understanding by revealing the intricate dialogue between cosmic dust and the radiant stars, creating a canvas upon which stellar narratives are inscribed. Our journey into stellar linguistics also leads us to contemplate the significance of our solar system's central luminary, the Sun, as we endeavor to comprehend its language and influence on Earth. These intertwined facets form the basis for engaging in a comprehensive investigation of stellar linguistics, paving the way for a deeper appreciation of the profound conversations occurring throughout the cosmos.

Historical Perspectives on Astronomical Symbols

Throughout history, human civilization has looked to the stars with awe and reverence, recognizing patterns and shapes in the celestial expanse above. Cultures from around the world have woven narratives around these stellar arrangements, giving rise to a tapestry of myths and legends that reflect their understanding of the universe. The ancient Greeks, for instance, depicted their gods and heroes as constellations, immortalizing their stories among the stars. Similarly, the indigenous cultures of Australia, the Americas, and Africa devised their own intricate star maps, linking them to creation myths and spiritual beliefs. These historical perspectives offer invaluable insights into how societies have interpreted and incorporated astronomical symbols into their belief systems and daily lives. Furthermore, the study of historical astronomical symbolism reveals the universal human impulse to seek meaning and understanding in the night sky. It underscores our innate connection to the cosmos, transcending cultural and temporal boundaries. By delving into the rich tapestry of historical astronomical symbolism, we gain a deeper appreciation for the enduring human quest to comprehend the celestial realm and the profound impact it has had on shaping diverse cultures across the ages.

Decoding the Celestial Alphabet: Constellations and Their Meanings

The study of celestial bodies has always captivated human imagination, leading to the creation of constellations and the attribution of mythological narratives and meanings to patterns in the night sky. These constellations serve as a symbolic language that has been passed down through generations, transcending cultural boundaries and embodying enduring significance. Each constellation represents a unique story deeply entrenched in cultural heritage, serving as a celestial alphabet infused with cosmic wisdom. Through centuries of observation and interpretation, these star clusters have become an integral part of our collective linguistic and cultural tapestry.

One example is the constellation Orion, which holds significant meaning across diverse cultures. In Greek mythology, Orion was a legendary hunter who was ultimately immortalized as a constellation after his death. To the ancient Egyptians, Orion's alignment with the Nile River symbolized rebirth and renewal. Similarly, indigenous cultures across the world have imparted their own interpretations onto the night sky, creating a rich mosaic of stories that continue to shape our understanding of the universe.

Furthermore, the scientific study of constellations has unveiled new layers of meaning. Modern astronomy has identified intricate patterns within these formations, allowing researchers to decode the cosmic symbols that exist beyond conventional folklore. For instance, the constellation Scorpius is not merely a scorpion from ancient sagas but also a representation of complex astrophysical phenomena. By deciphering the nuances embedded in these celestial arrangements, scientists are gaining profound insights into the composition, lifecycle, and behavior of celestial bodies, thereby expanding our cosmic lexicon.

This study is not limited to a single cultural or temporal context. It encompasses a holistic approach that integrates ancient wisdom with contemporary scientific inquiry, providing a comprehensive framework to understand the profound language written across the heavens. As we delve deeper into the intricacies of celestial symbolism, we venture into an ever-expanding realm of knowledge where the enigmatic beauty of the cosmos converges with the meaningful narratives woven by humanity.

Astrophysical Phenomena as Linguistic Constructs

Our exploration of the language of the stars takes us beyond the mere observation of constellations and celestial bodies. It leads us to an intriguing realization: astrophysical phenomena serve as complex linguistic constructs that convey vital information about the cosmos. The elegant dance of galaxies, the violent fireworks of supernovae, and the cosmic symphonies of black holes are not just awe-inspiring spectacles; they are also rich sources of information encoded in the fabric of the universe.

Consider the graceful spirals of distant galaxies, each swirl and bend telling a story of cosmic evolution. The language of these galactic structures reveals a narrative of gravitational interactions, star formation, and the ebb and flow of stellar life cycles. This grand celestial calligraphy inscribes the saga of billions of years, offering us a glimpse into the intricate web of cosmic history.

Moving closer to home, the life cycle of stars unfolds as a saga of birth, maturation, and eventual demise, each stage characterized by distinctive emissions that form an astronomical lexicon. Spectroscopy, the art of dissecting starlight into its constituent wavelengths, allows astronomers to decipher this lexicon and reveal the chemical composition, temperature, and motion of distant celestial bodies. The delicate ballet of spectral lines unveils the elemental vocabulary of stars and galaxies, akin to deciphering the ancient runes of a distant civilization.

Furthermore, we encounter the cosmic semantics embedded within the interstellar medium. Dust clouds, gas nebulae, and energetic particle streams compose a unique dialect of cosmic semiotics. These enigmatic forms of matter serve as carriers of information, shaping the dynamics of star formation and astronomical events. Their interaction with radiation produces a tapestry of signals, akin to an ethereal language waiting to be interpreted by the curious minds of astrophysicists and cosmologists.

Astrophysical phenomena also manifest as celestial timekeepers and messengers of cosmic epochs. Pulsars, the highly magnetized remnants of massive stars, act as galactic metronomes, emitting regular pulses of electromagnetic radiation. Their rhythmic pulsations carry insights into the nature of compact stellar remnants and provide a cosmic chronicle unraveling the passage of time in the universe. Meanwhile, quasars, luminous beacons powered by supermassive black holes, serve as signposts to distant cosmic eras, narrating the tumultuous youth of the cosmos.

In unraveling the linguistic tapestry woven by astrophysical phenomena, we venture into the heart of cosmic communication. From the cosmic ballet of galactic whirls to the radiant verses penned by the stars, the universe's language transcends the boundaries of spoken words, inviting us to explore its profound narratives and decode the secrets inscribed in the celestial script.

Analyzing Starlight: Spectroscopy and Semiotics

Spectroscopy, the study of the interaction between matter and radiated energy, has become an invaluable tool in decoding the language of the stars. By analyzing the light emitted or absorbed by celestial bodies, astronomers have been able to unravel a wealth of information about the composition, temperature, and motion of distant objects in the cosmos. Through the lens of spectroscopy, starlight reveals a complex narrative, rich with insights into the physical and chemical properties of celestial bodies.

Semiotics, the study of signs and symbols, finds a natural application in the realm of astrophysics. The light we receive from stars carries a multitude of encoded messages, waiting to be deciphered. Each element present in a star's atmosphere leaves its signature in the spectrum of light it emits, akin to a unique linguistic character forming part of an intricate cosmic alphabet. By meticulously examining these spectral lines, astrophysicists can discern the elemental makeup of a star, unveiling the story of its genesis and evolution.

Interstellar Medium: Messages in Cosmic Dust

The interstellar medium, the vast expanse of space between stars and galaxies, harbors intricate messages within its cosmic dust. These seemingly ephemeral particles hold profound significance as carriers of information about the physical and chemical history of the universe. Comprised of a diverse array of molecules and grains, interstellar dust serves as a silent but compelling storyteller, offering glimpses into the processes that have shaped the cosmos over immense spans of time.

These cosmic particles not only bear witness to the astrophysical events that have occurred within their vicinity but also play a crucial role in the formation of stars and planetary systems. As light traverses the interstellar medium, it interacts with the myriad components of cosmic dust, leading to a phenomenon known as extinction. This absorption and scattering of light provide researchers with invaluable insights into the composition and distribution of interstellar dust, unlocking the secrets contained within this enigmatic medium.

Moreover, recent advances in observational techniques and theoretical models have enabled scientists to delve deeper into the role of interstellar dust as a medium for interstellar chemistry. The formation of complex organic molecules within these cosmic grains presents a compelling narrative of the chemical evolution of the universe, offering profound implications for our understanding of the origins of life and the potential for habitable environments beyond Earth.

Furthermore, the study of interstellar dust holds immense promise for unraveling the mysteries of interstellar communication. It has been postulated that cosmic dust particles could serve as carriers for interstellar messages, potentially facilitating interactions between distant civilizations or cosmic entities. With further exploration and refinement of these theories, the prospect of decoding and interpreting messages within the interstellar medium represents a tantalizing avenue for the advancement of exo-languages and the broader field of astro-linguistics.

In essence, delving into the realm of interstellar dust unveils a rich tapestry of scientific inquiry and philosophical contemplation, presenting profound questions about the nature of the cosmos and our place within it. As we continue to explore and decipher the messages within this cosmic medium, we are poised to encounter transformative revelations that may redefine our perceptions of existence and the universal dialogue that transcends the boundaries of space and time.

Pulsars and Quasars: Beacons of the Universe

Pulsars and quasars stand as celestial beacons, emitting signals that traverse unfathomable distances to impart critical insights into the language of the cosmos. Pulsars are highly magnetized, rotating neutron stars that emit beams of electromagnetic radiation. These rapidly spinning remnants of massive stars serve as cosmic timekeepers, their periodic signals akin to the rhythmic beats of the universe's heart. By meticulously deciphering the temporal patterns of pulsar emissions, astronomers glean invaluable knowledge about the fundamental properties of spacetime and the dense matter within these enigmatic entities.

In stark contrast, quasars captivate the astronomical community with their intense luminosity and profound implications for our understanding of galactic dynamics. These quasi-stellar radio sources represent the energetic cores of distant galaxies, often harboring supermassive black holes that fuel their radiant displays. Emitting copious amounts of electromagnetic energy across the entire spectrum, quasars illuminate the intricate dance of matter as it plunges into the gravitational abyss of colossal black holes. By probing the spectral signatures of quasar emissions, scientists probe the cosmic history of galaxy formation and the evolution of the universe itself.

It is the unique and precisely orchestrated characteristics of pulsars and quasars that afford them the distinction of being celestial beacons—lighthouses in the darkness of space. The regularity of pulsar signals and the dazzling brilliance of quasar emissions serve as testaments to the profound interplay between matter, energy, and the fabric of the cosmos. As humanity endeavors to unravel the mysteries of the universe, pulsars and quasars stand as remarkable signposts, guiding us toward a deeper comprehension of the cosmic symphony that surrounds us.

The Sun: Understanding Our Closest Star

Our sun, a massive sphere of seething hot gas, has captivated human curiosity for millennia. As the center of our solar system, the sun holds profound significance in both scientific and cultural realms. Understanding the complexities and nuances of this celestial body is crucial for comprehending the broader dynamics of the cosmos and its impact on life on Earth. At its core, the sun's immense pressure and temperature fuel nuclear fusion processes, generating an unfathomable amount of energy that radiates into space.

The sun's energy not only sustains life on our planet but also exerts a profound influence on its atmospheric dynamics and climate patterns. Solar irradiance directly contributes to Earth's overall energy budget, shaping weather systems, ocean currents, and ecological cycles. Moreover, fluctuations in solar activity can have far-reaching implications, influencing everything from agricultural productivity to technological infrastructure.

From a scientific perspective, studying the sun presents an array of challenges and opportunities. Advanced telescopic observations, coupled with space-based missions, have provided unprecedented insights into solar phenomena such as sunspots, solar flares, and coronal mass ejections.

The intricate interplay of magnetic fields within the sun's atmosphere gives rise to these mesmerizing yet potentially disruptive events, shedding light on the complex dynamics inherent to stellar bodies.

Furthermore, delving into solar physics enables us to unravel fundamental principles governing stars across the universe. Comparative analyses between our sun and other stellar entities offer valuable benchmarks for understanding the life cycles, energy production mechanisms, and evolutionary pathways of stars throughout the cosmos. This interdisciplinary approach not only contributes to astronomical research but also fosters a deeper appreciation for the interconnectedness of celestial bodies within the cosmic tapestry.

Beyond its scientific significance, the sun holds a prominent place in myriad cultures and belief systems. Across diverse civilizations, the sun has been revered as a symbol of life, vitality, and renewal. From ancient mythologies to modern cultural expressions, the sun's enduring presence has woven itself into art, religion, and folklore, serving as a timeless source of inspiration and contemplation.

As we seek to grasp the intricacies of our closest star, the sun stands as a testament to the marvels of astrophysical inquiry and the evocative power of celestial symbolism. Each new revelation about the sun enriches our understanding of cosmic phenomena and reaffirms the indispensable role of astrophysics in illuminating the mysteries of the universe.

From Mythology to Modern Science: Cultural Interpretations of the Stars

Throughout history, civilizations across the globe have looked up to the shimmering night sky and woven captivating tales and myths around the celestial bodies they observed. These stories, rooted in cultural traditions and beliefs, formed the cornerstone of early astronomy and helped lay the foundation for humanity's understanding of the cosmos. From the Greek legend of Orion the Hunter to the Hindu mythological deity Surya, the divergent narratives surrounding the stars reflect the rich tapestry of human imagination and ingenuity. As societies evolved and scientific knowledge advanced, these mythological interpretations gradually merged with empirical observations, sparking a monumental shift towards modern scientific inquiry. The transition from folklore to astrophysics represents an intriguing crossover where cultural symbolism intersects with empirical data. This fusion not only embellished the field of astronomy with captivating narratives but also provided a platform for cross-cultural dialogue and exchange of ideas. Through the ages, constellations and celestial phenomena have been intertwined with religious rituals, agricultural practices, and navigational techniques, playing a vital role in shaping various aspects of human societies. Moreover, by tracing the evolution of celestial interpretations, researchers can gain invaluable insights into the collective consciousness and worldview of diverse cultures. In contemporary society, the interplay between mythology and astronomy continues to inspire awe and fascination, serving as a bridge between the realms of science and spirituality. Acknowledging the cultural significance of stars and embracing the interdisciplinary nature of astronomical studies allows for a more holistic and inclusive approach to exploring the universe. By integrating historical narratives with rigorous scientific methodologies, we can unravel the complexities of human perception while shedding light on the timeless allure of the cosmic theater.

Future Directions in Exo-Linguistics

The future of exo-linguistics holds boundless possibilities, offering tantalizing prospects for unraveling the language of the universe beyond our own celestial sphere. As technological advancements continue to push the boundaries of our understanding, the study of cosmic communication stands at the threshold of unprecedented breakthroughs.

One of the most promising avenues for future research lies in the exploration of extraterrestrial radio signals. The continued development of radio telescopes and the emergence of next-generation instruments such as the Square Kilometer Array (SKA) promise to revolutionize our ability to detect and analyze signals from distant civilizations. Coupled with innovative signal processing algorithms and machine learning techniques, these advancements hold the potential to decode complex extraterrestrial transmissions that have eluded us thus far.

Moreover, the advent of space missions focused on exoplanetary exploration presents an unparalleled opportunity to directly study alien environments and possibly encounter forms of non-human intelligence. By leveraging advanced spectroscopic analysis and remote sensing technologies, scientists aim to identify biosignatures and undertake in-depth studies of exoplanetary atmospheres and surfaces to uncover potential linguistic artifacts or environmental cues indicative of intelligent life.

In addition, interdisciplinary collaborations at the intersection of astrophysics, linguistics, and information theory will play a pivotal role in shaping the trajectory of exo-linguistic research. The synthesis of expertise across diverse fields will engender novel methodologies and analytical frameworks for understanding and interpreting potential interstellar communications. Furthermore, the application of artificial intelligence and computational linguistics in sifting through vast datasets obtained from space-based observatories opens new vistas for identifying patterns or anomalies that may signify deliberate attempts at cosmic discourse.

As humanity sets its sights on ambitious ventures such as manned missions to Mars and the establishment of lunar outposts, the need to develop protocols for interstellar communication becomes increasingly pressing. Ethical considerations and the implications of potential contact with extraterrestrial intelligences command thoughtful examination, prompting the formulation of protocols to guide future interactions. These endeavors underscore the urgency of fostering international cooperation and instituting universal standards for responsible interstellar messaging to mitigate inadvertent misunderstandings or conflicts stemming from misinterpreted signals.

Ultimately, the pursuit of exo-linguistics represents a profound intellectual odyssey, an undertaking that spans disciplines and fuels the imagination with the prospect of engaging in dialogue with civilizations far beyond our terrestrial confines. While the path ahead is replete with challenges and uncertainties, the allure of comprehending the cosmic lexicon binds humanity in a collective quest to decipher the enigmatic languages of the stars.

Chapter 7

Vegetation Vocabularies: Understanding Plant Communications

Plant Communication

Plant communication is a fascinating and intricate field of botanical science that delves into the complex ways in which plants interact and signal with each other and the environment. This chapter aims to provide an exploration of the introductory concepts central to understanding plant communication, defining its scope within the broader context of ecological interactions. As we embark on this journey, it is vital to bear in mind the interconnectedness of all living organisms and the immense value in deciphering the language of the natural world. In linking back to the previous chapter on star communication, where we explored celestial languages and the mysteries of cosmic signals, we draw parallels to the nuanced communication systems found on Earth. The study of plant communication offers profound insights into the resilience and adaptability of living organisms, and highlights the intricate intelligence embedded in the natural world.

The Basics of Botanical Signaling

Plants, often underestimated in their ability to communicate, possess a remarkably complex system of signaling that allows them to respond to environmental stimuli and interact with other organisms. At the core of botanical signaling is the intricate network of chemical messengers, receptors, and pathways that enable plants to convey and receive vital information. The process of botanical signaling begins with the detection of external stimuli, such as changes in temperature, light intensity, or the presence of nutrients and potential threats. Upon sensing these cues, plants activate signaling pathways that trigger specific responses, such as growth, defense mechanisms, or reproductive strategies. These responses are carefully orchestrated through a series of molecular events that involve the production and transmission of chemical signals. Plants utilize an array of compounds, including hormones, volatile organic compounds (VOCs), and secondary metabolites, to communicate with neighboring plants, beneficial microbes, and other organisms within their ecosystem. The exchange of these chemical signals not only facilitates inter-plant communication and mutualistic interactions but also plays a crucial role in defending against herbivores, pathogens, and abiotic stressors. Furthermore, botanical signaling is not limited to chemical communication; it also encompasses physical and mechanical signals. For instance, plants can transmit and receive mechanical stimuli, such as touch or vibration, which can elicit defensive responses or influence growth patterns. The intricate interconnectedness of botanical signaling highlights the sophisticated and adaptive nature of plant communication, underscoring the pivotal role it plays in ecological dynamics and evolutionary processes. Understanding the basics of botanical signaling provides a deeper insight into the complex language of plants and unveils the remarkable sophistication of their interactions with the environment. This foundational knowledge serves as a springboard for exploring the multifaceted facets of plant communication and its far-reaching implications across diverse fields, from agriculture and ecology to biotechnology and conservation.

Chemical Messengers in Plant Interactions

Plants, unlike animals, are rooted in the ground and cannot move to escape danger or seek out advantageous environments. Instead, they have evolved a sophisticated and intricate system of chemical messaging to interact with their surroundings. This system allows them to communicate with other plants, defend against predators, and respond to environmental changes. At the heart of this botanical communication network are the chemical messengers that convey vital information from one plant to another. Plants release a diverse array of volatile organic compounds (VOCs) into the air, which serve as long-distance signals to neighboring plants. These VOCs can act as warnings in response to herbivore attacks, alerting nearby plants to prepare their own defense mechanisms. In addition to airborne communication, plants also utilize chemical signaling through their root systems. Through the release of specific compounds, plants can influence the microbial communities in the soil, forming symbiotic relationships with beneficial organisms or deterring harmful pathogens. Furthermore, chemical messengers play a crucial role in mediating plant interactions with other organisms, such as attracting pollinators or repelling herbivores. The complexity of these chemical conversations is astounding, with plants able to detect and respond to minute quantities of specific compounds. Studies have revealed that some plants can even recognize and differentiate between the chemical profiles of different neighboring species, allowing for tailored responses to potential threats. The significance of chemical messengers in plant interactions stretches beyond individual plants, influencing entire ecosystems and shaping ecological dynamics. By understanding the intricacies of these botanical communications, we gain valuable insights into the interconnectedness of the natural world and can explore innovative approaches for sustainable agriculture, ecosystem management, and conservation efforts.

Electrical Signals and Their Implications

Plants communicate not only through chemical signals but also via electrical signals, a form of communication that has profound implications for our understanding of plant behavior and interactions with their environment. The transmission of electrical signals in plants occurs through long-distance signaling networks, enabling them to respond to various environmental stimuli and coordinate their growth and development. One of the key components in electrical signaling is membrane potential, which refers to the difference in electric charge between the inside and outside of the plant cells. This membrane potential allows for rapid transmission of electrical signals across the plant's tissues and organs. Another crucial aspect of electrical signaling in plants is action potentials, which are rapid changes in membrane potential that play a role in transmitting information over long distances within the plant. These action potentials can be triggered by various stimuli, including mechanical disturbances, light, and pathogen attacks. The propagation of electrical signals in plants has far-reaching implications for their responses to stress and defense mechanisms. For instance, when a plant experiences herbivory or physical damage, it can rapidly transmit electrical signals to trigger defensive responses such as the release of toxic compounds or the production of physical barriers. Furthermore, electrical signals are involved in coordinating plant responses to environmental factors such as light, temperature, and water availability, playing a critical role in the regulation of processes such as photosynthesis, transpiration, and stomatal movement. Understanding the intricacies of plant electrical signaling is essential for developing innovative agricultural practices and pinpointing strategies to enhance crop resilience in the face of environmental challenges. In addition, delving into the world of plant electrical communication opens up new possibilities for bio-inspired technologies and materials. By drawing inspiration from the sophisticated electrical signaling networks in plants, researchers have the potential to develop novel bioelectronic devices and sensors that mimic the efficiency and adaptability of plant-based communication systems.

The study of plant electrical signaling represents a fascinating intersection of biology, physics, and engineering, offering fruitful avenues for further exploration and innovation in the realm of plant science and technology.

Root Networks: The Underground Language

Root networks form the cornerstone of plant communication, constituting the underground language through which plants interact and support each other within their ecosystem. In this subterranean realm, an intricate web of mycorrhizal fungi connects the roots of various plants, enabling them to exchange nutrients, water, and chemical signals. This symbiotic relationship, known as mycorrhizal network, allows plants to communicate and provide assistance to one another, establishing a complex support system beneath the surface. Through these fungal highways, plants can transmit warning signals in response to threats such as herbivore attacks or diseases, effectively activating defense mechanisms in neighboring plants to enhance their resistance. Furthermore, the mycorrhizal network serves as a conduit for resource sharing and distribution, fostering a collaborative environment where plants cooperate to thrive mutually. Besides mycorrhizal interactions, roots also release volatile compounds that convey information to neighboring plants, notifying them about their presence and potentially influencing their growth patterns. This phenomenon, known as allelopathy, showcases how root systems participate in dialogues aimed at achieving ecological balance and cooperation. Additionally, recent research has unveiled the role of root exudates in shaping microbial communities and modulating soil chemistry, further highlighting the pivotal role of roots in underground communication networks. Understanding the intricacies of root networks not only sheds light on the interconnectedness of plant life but also offers insights into potential applications in agriculture, forestry, and environmental conservation. By delving into the underground language of root networks, scientists are uncovering innovative strategies for optimizing crop health, promoting sustainable land management practices, and preserving the delicate balance of natural ecosystems.

Phytoacoustic Signals: Sound within Silence

Plants have long been perceived as silent, motionless entities, yet recent research has unveiled the astonishing world of phytoacoustic signals – the communication of plants through sound. Phytoacoustic signaling involves the emission of acoustic waves, vibrations, and subtle frequencies, often imperceptible to human hearing. These signals play a pivotal role in orchestrating complex interactions within the plant kingdom and beyond.

At its core, phytoacoustic signaling allows plants to convey vital information to one another, such as impending environmental stress, threats from herbivores, or the presence of beneficial organisms. Through emitting and receiving acoustic cues, plants can initiate defensive mechanisms, alter their biochemical composition, and coordinate collective responses to external stimuli. This form of communication not only enhances individual plant survival but also fosters symbiotic relationships and ecological balance within diverse ecosystems.

Furthermore, the study of phytoacoustic signals has significant implications for agriculture and horticulture. By deciphering the auditory language of plants, researchers and cultivators can gain insights into optimizing crop growth, detecting pest infestations, and promoting sustainable agricultural practices. Harnessing this knowledge may revolutionize farming techniques, leading to more efficient and environmentally conscious approaches to food production.

Intriguingly, phytoacoustic signals extend beyond the confines of the botanical realm, influencing various organisms within their acoustic domain. Research suggests that certain insect species are attuned to these subtle vegetative communications, further illustrating the interconnectedness of terrestrial life through sound. Additionally, the potential application of phytoacoustic knowledge in bio-inspired technologies and sound-based therapies holds promise for diverse fields, from biomimicry to holistic health practices.

While the study of phytoacoustic signals is still in its infancy, the burgeoning insights into this silent symphony of the plant world present boundless opportunities for exploration and innovation. As we continue to unravel the complexities of phytoacoustic communication, we unveil the hidden melodies of nature, enriching our understanding of the intricate web of life on Earth.

Photographic Responses: Light as a Communicative Medium

In the intricate web of plant communication, light plays a pivotal role as a potent medium for conveying information and signaling. Plants have evolved remarkable mechanisms to sense and respond to light, utilizing it not only for photosynthesis but also as a means of communication. Photographic responses in plants encompass a diverse range of reactions to varied light stimuli, shedding light on the complexity of their interaction with the environment. When exposed to light, plants undergo a series of responses that are crucial for their growth, development, and adaptation. The phenomenon of phototropism, where plants bend towards a light source, showcases their ability to perceive and orient themselves in response to light. Furthermore, the regulation of flowering, seed germination, and various developmental processes are intricately linked to light signals, emphasizing its significance as a communicative cue in the plant kingdom. Beyond these visible manifestations, plants also harness light signals to communicate with each other and neighboring organisms. From the release of volatile organic compounds to attract pollinators to the production of chemical defenses in response to specific light wavelengths, plants adeptly utilize light as a language to interact with their surroundings. Delving deeper into this botanical lexicon, researchers continue to unravel the nuanced ways in which plants employ light-mediated communication to navigate their ecosystem and orchestrate complex ecological relationships. Moreover, the study of photomorphogenesis has revealed the profound impact of light on shaping plant form and function, highlighting the far-reaching implications of light as a communicative medium in the botanical realm. As we strive to comprehend the intricate interplay between plants and light, it becomes evident that unlocking the secrets of photographic responses not only enriches our understanding of plant biology but also offers potential insights for harnessing sustainable agricultural practices and environmental conservation. By delving into the captivating realm of photographic responses, we gain a profound appreciation for the profound role of light as a communicative medium that shapes the vibrant tapestry of life on our planet.

Inter-species Communication: How Plants Talk with Other Organisms

Plants are not solitary entities dwelling in silence; they actively engage in intricate communication with a myriad of organisms that share their environment. Through a multifaceted repertoire of chemical, acoustic, and visual signals, plants interact with insects, animals, fungi, and even microorganisms in an elaborate dance of mutual influence. The plant kingdom has evolved an array of mechanisms to attract beneficial partners, deter or repel potential threats, and establish complex symbiotic relationships. At the core of this inter-species communication lies the emission of volatile organic compounds that serve as olfactory signals, attracting specific pollinators or repelling herbivores. Additionally, plants utilize visual cues such as floral coloration and patterns to entice pollinators, ensuring successful reproduction. Moreover, these visual signals may also act as warning signs to potential predators, conveying the presence of toxic compounds or anti-nutrient defenses within the plant's tissues. In the underground realm, plants exchange biochemical messages through their root systems, forming intricate networks of communication with neighboring individuals and symbiotic fungi. Mycorrhizal associations, for instance, epitomize the cooperative dialogue between plants and fungi, facilitating nutrient exchange and enhancing resilience against environmental stress. Furthermore, recent research has unveiled the phenomenon of phytoacoustic signaling, wherein plants emit ultrasonic vibrations triggered by internal physiological processes or external stimuli. These acoustic emissions may serve as a means of defense against herbivores, as well as facilitate communication with other organisms. Beyond these direct modes of communication, plants also induce indirect responses in neighboring species through the modification of their immediate environment. By releasing secondary metabolites into the soil or altering the chemical composition of the air, plants can influence the behavior and growth of nearby organisms. The interconnected web of interactions between plants and other living beings highlights the profound implications of these communications on ecosystem dynamics and biodiversity. Understanding the intricacies of inter-species communication in the botanical world unveils a tapestry of coexistence and reciprocity, enriching our perspective on the subtle yet pervasive interplay of life forms within natural habitats.

Impacts of Environmental Stress on Plant Communications

Environmental stress poses a significant challenge to the intricate network of plant communication. It is imperative to understand how various environmental factors such as extreme temperatures, changes in precipitation patterns, soil salinity, air pollution, and habitat destruction affect the ability of plants to effectively communicate and interact with their surroundings. The impact of these stressors can be observed at various levels of plant communication networks.

One of the key consequences of environmental stress is the alteration of volatile organic compounds (VOCs) emitted by plants. These VOCs play a crucial role in signaling and defense mechanisms, but under stress, their composition and quantity may change, affecting the overall chemical communication within and between plant species. Furthermore, exposure to pollutants can disrupt the ability of plants to produce and process these volatile compounds, leading to complications in intra- and inter-species interaction dynamics.

In addition to chemical signaling, environmental stress can also influence electrical signaling in plants. Studies have shown that fluctuations in environmental conditions such as drought or nutrient deficiency can impact the transmission of electrical signals within plant tissues. This disruption in electrical communication pathways can hinder important physiological responses and coordination among different parts of the plant, potentially compromising its resilience and adaptive capacity.

The intricate mycorrhizal networks that facilitate below-ground information exchange are also vulnerable to environmental stress. Mycorrhizal associations are essential for nutrient uptake and communication among plants, enhancing their ability to withstand environmental challenges. However, disturbances in soil structure and composition resulting from stressors like deforestation, urbanization, or agricultural practices can disrupt these symbiotic relationships, limiting the effective transfer of resources and information among interconnected plants.

Moreover, environmental stress can lead to alterations in plant growth patterns, flowering cycles, and seed production, all of which have implications for the timing and content of communicative signals. As plants adapt to changing environmental conditions, their phenological cues and behavioral responses may shift, influencing the temporal and spatial dynamics of their communication strategies.

Understanding the impacts of environmental stress on plant communications is essential for devising strategies to mitigate its negative effects and enhance the resilience of plant communities. By delving into the intricacies of how stressors modulate the language of vegetation, researchers can contribute to the development of more sustainable agricultural practices, conservation efforts, and restoration initiatives geared towards preserving the vital relationships forged through botanical communication.

Future Perspectives in Botanical Communication Research

As our understanding of botanical communication continues to expand, the future of research in this field holds promise for groundbreaking discoveries and practical applications. One area of great interest lies in exploring the potential for harnessing plant communication for sustainable agriculture and environmental conservation. By delving deeper into how plants signal distress, attract beneficial organisms, and communicate with neighboring species, researchers can potentially develop innovative strategies to enhance crop resilience, reduce dependence on chemical interventions, and foster biodiversity in agricultural landscapes. Furthermore, the study of plant communication offers opportunities to gain insights into ecosystem dynamics and develop ecologically sensitive management practices. This interdisciplinary approach has the potential to revolutionize our approaches to farming, land use, and conservation, paving the way for more harmonious coexistence between human activities and natural environments. Another exciting avenue for future research involves leveraging advances in technology to unravel the intricacies of plant signaling. Integrating cutting-edge tools such as advanced imaging techniques, molecular analysis, and bioinformatics can provide unprecedented detail into the mechanisms and nuances of botanical communication. This technological leap not only enhances our fundamental understanding of plant signaling but also presents opportunities for developing innovative applications in areas such as precision agriculture, biomimicry, and pharmaceutical discovery. Furthermore, the potential for interdisciplinary collaboration with fields such as materials science and engineering opens up new frontiers in biomaterials and bioinspired technologies, where insights from plant communication could inform the design of novel functional materials and systems. As research in plant communication progresses, one critical direction for the future involves considering the ethical implications and responsibilities associated with this knowledge. Understanding the complexities of plant communication prompts contemplation on the ethical treatment of plant life and the broader implications for how we interact with the natural world. It calls for a thoughtful examination of the ethical considerations in areas such as biotechnology, agriculture, and ecological management. Moreover, it necessitates reflection on the societal attitudes towards plant life and the realization of the interconnectedness of all living organisms. Recognizing these ethical dimensions will be crucial as we navigate the applications and potential impacts of advancements in botanical communication research. In conclusion, the future of botanical communication research is teeming with opportunities for paradigm-shifting discoveries and transformative applications. By exploring the intersections of plant signaling with agriculture, technology, and ethics, researchers are poised to unlock new realms of understanding and cultivate a more holistic relationship with the natural world.

Chapter 8

The Language of the Skies

Sky Communication

Historical observations of celestial patterns have long fascinated and perplexed humankind, as they hold profound significance in shaping our understanding of the world around us. Observing the skies has been integral to various cultures and civilizations throughout history, with many ancient societies attributing divine or mystical meanings to celestial phenomena. However, beyond mere fascination, these observations have provided invaluable insights into the language of the skies, giving rise to the discipline of atmospheric communication. Atmospheric communication encompasses the study and interpretation of celestial patterns, meteorological phenomena, and atmospheric dynamics to decipher the messages conveyed by the sky. It represents a sophisticated form of communication that carries essential information about weather patterns, climatic trends, atmospheric conditions, and broader environmental interactions. The systematic analysis of celestial patterns allows us to glean vital knowledge about Earth's interconnected systems and their influences on our planet's ecology, agriculture, and human activities. Exploring the basics of atmospheric communication unveils compelling correlations between celestial events and natural processes, shedding light on the intricate interplay between the skies and our terrestrial environment. By delving into this realm of study, we embark on a journey to comprehend the profound influence of celestial patterns on the Earth's ecosystems, cultural beliefs, and historical narratives. Through meticulous examination and interpretation, we can begin to decode the intricate language of the skies, fostering a deeper appreciation for the intricate tapestry of nature's communications above us.

Historical Observations of Celestial Patterns

Throughout history, civilizations have observed and recorded celestial patterns, recognizing their significance in navigating the natural world. Ancient cultures looked to the skies for guidance and knowledge, developing intricate systems to interpret celestial movements and formations. The Mesopotamians meticulously documented the paths of stars and planets, finding correlations with seasonal changes and agricultural cycles. Their observations laid the foundation for astrology and informed the development of early calendars. Similarly, the ancient Egyptians closely monitored the movements of celestial bodies, linking them to religious beliefs and rituals. The alignment of certain stars and constellations with specific events served as a fundamental aspect of their cosmology and societal structure. In China, the study of celestial patterns led to the creation of the lunisolar calendar, which harmonized solar and lunar cycles, guiding agricultural activities and cultural festivals. The Mayans, renowned for their advanced understanding of astronomy, constructed elaborate observatories to track celestial phenomena. Their precise astronomical calculations allowed them to predict celestial events and organize their ceremonial calendar. Closer to home, Native American tribes correlated celestial occurrences with natural rhythms, incorporating these observations into their traditions and ceremonies. The expansive cultural diversity of historical sky observations demonstrates the inherent human desire to comprehend and harness the language of the skies. These ancient interpretations provide valuable insights into the deep-rooted significance of celestial patterns across various

societies, offering a timeless perspective on our relationship with the cosmos
.

Meteorological Symbols and Signs

The study of meteorological symbols and signs provides a fascinating insight into the language of the skies. Throughout history, different cultures have observed and interpreted atmospheric phenomena as symbolic messages or omens from the heavens. From simple observations of cloud formations to complex interpretations of celestial events, meteorological symbols and signs have played a significant role in human understanding and interaction with the natural world. The ability to discern the meanings behind these signs has contributed to both practical weather prediction and spiritual or cultural significance.

Clouds, for example, have long been regarded as indicators of impending weather patterns. Their appearance, movement, and formations provide valuable information about atmospheric conditions. Cumulonimbus clouds, for instance, are associated with thunderstorms, while cirrus clouds may signal approaching changes in the weather. Moreover, ancient civilizations often linked certain cloud shapes to mythical creatures and symbols, integrating their interpretations into stories and belief systems. This blending of science and mythology demonstrates the profound impact of meteorological symbols on human culture.

Beyond cloud formations, celestial events such as solar and lunar eclipses, comets, and meteor showers have historically been perceived as portents of future events. The precise timing and nature of these phenomena were seen as messages from the cosmos, often inspiring awe and fear, and leading to predictions of wars, catastrophes, or bountiful harvests. Even today, some communities continue to attach special significance to celestial occurrences, preserving the ancient practices of interpreting such ev ents.

Understanding meteorological symbols and signs is crucial not only for historical and cultural reasons but also for practical purposes. Modern meteorology, heavily reliant on scientific data and technology, continues to benefit from an awareness of traditional interpretations and observations. By integrating historical wisdom with contemporary knowledge, meteorologists can refine their understanding of atmospheric phenomena and improve weather forecasts. Additionally, the cultural significance of these symbols enriches our appreciation of the natural world and helps us connect with our ancestors' relationship to the skies. In essence, the study of meteorological symbols and signs serves as a bridge between the scientific and cultural dimensions of our understanding of the skies.

The Role of Cloud Formations

Cloud formations are not just random patterns in the sky; they are a crucial part of nature's intricate language. Each cloud formation carries its own unique message, communicating information about the atmospheric conditions and impending weather events. Understanding the role of cloud formations is essential for deciphering the language of the skies. Clouds can be classified into several main types, including cirrus, cumulus, stratus, and nimbus clouds, each with distinct characteristics and meanings. Cirrus clouds, for instance, are wispy and featherlike in appearance, often signaling fair weather, while cumulonimbus clouds, towering and dense, may foretell thunderstorms or heavy rainfall. It is through careful observation and interpretation of these formations that we gain insights into the ever-changing dynamics of the atmosphere. Additionally, the altitude and vertical development of clouds provide valuable information about atmospheric stability and moisture content. For example, lenticular clouds forming near mountain ranges indicate strong winds and turbulence at high altitudes, offering insight into the potential for aviation hazards. Moreover, certain cloud formations such as mammatus clouds indicate severe turbulence or stormy conditions in the vicinity. Understanding the role of cloud formations also involves recognizing the impact of human activities on cloud formation and precipitation processes. Anthropogenic influences, such as air pollution and industrial emissions, can lead to the formation of artificial clouds known as contrails, affecting local weather patterns and climate dynamics. Moreover, advancements in technology have enabled scientists to study the microphysical properties of clouds, shedding light on their role in global climate systems and the hydrological cycle. Through satellite imagery and remote sensing techniques, researchers can analyze the composition and behavior of cloud formations on a global scale, contributing to our understanding of Earth's complex atmospheric processes. In essence, the study of cloud formations extends beyond mere meteorological observation; it represents a multidisciplinary endeavor encompassing atmospheric science, environmental studies, and climatology, offering profound insights into the interconnectedness of Eart h's natural systems.

Wind Patterns and Their Messages

Wind, an essential component of the Earth's atmospheric system, holds a wealth of communicative capabilities. In the realm of natural linguistics, wind patterns play a significant role in conveying messages from the skies to the Earth. The intricate dance of air currents carries with it stories of distant lands, upcoming weather changes, and even ecological shifts. From gentle breezes to powerful gales, the language of the winds provides invaluable insights into the interconnectedness of our planet's ecosystems.

At a fundamental level, understanding wind patterns involves delving into the dynamics of atmospheric circulation. The movements of air masses across the globe create distinct wind patterns that hold valuable information regarding regional climates, weather events, and broader environmental processes. By decoding these patterns, researchers can discern the migratory routes of birds, the dispersal of seeds and pollen, and even the propagation of wildfires.

Furthermore, indigenous cultures and ancient civilizations have long revered the significance of wind patterns, recognizing them as carriers of omens, warnings, and blessings. The traditions of reading wind language, passed down through generations, offer profound wisdom in understanding the interplay between natural forces and human existence. These cultural perspectives shed light on the diverse ways in which wind patterns have been interpreted and integrated into various societies worldwide.

Moreover, the advent of modern technology has ushered in new avenues for comprehending the nuanced messages embedded within wind patterns. Advanced meteorological instruments, such as anemometers and weather balloons, enable scientists to gather precise data on wind speed, direction, and variability. Through meticulous observation and analysis, these tools aid in deciphering the intricate narratives woven into the fabric of the atmosphere, enhancing our ability to forecast weather phenomena and anticipate potential environmental impacts.

The implications of understanding wind patterns extend beyond terrestrial realms, encompassing the realms of aviation, maritime navigation, and renewable energy generation. Pilots harness wind data to optimize flight paths, while sailors rely on wind forecasts for safe passage across oceans. Additionally, wind energy technologies capitalize on the kinetic energy of air currents, offering sustainable solutions to power communities and industries. Unraveling the language of wind holds promise for enhancing - efficiency, safety, and sustainability across a spectrum of human activities.

In essence, wind patterns serve as eloquent messengers, speaking in the ethereal dialect of the skies. Their messages echo the harmonious choreography of Earth's systems, weaving tales of ecological harmony, seasonal transitions, and the timeless rhythms of nature. By deepening our comprehension of the subtle nuances encoded within wind patterns, we gain profound insights into the interconnected web of life on our planet.

Precipitation as a Language: Rain, Snow, and Hail

As we delve into the language of the skies, it becomes evident that precipitation is not merely a meteorological event, but a form of communication from the atmosphere to the earth. Each type of precipitation - rain, snow, and hail - carries its own distinct message, conveying valuable information about atmospheric conditions and patterns. Rain, the most common form of precipitation, speaks of renewal, fertility, and cleansing. Across different cultures and throughout history, rain has been revered as a life-giving force, essential for agricultural prosperity and spiritual rejuvenation. Its rhythmic patter against the earth's surface echoes a timeless language of rebirth and growth. Snow, with its ethereal beauty, conveys a sense of tranquility and purity. Its delicate crystalline structure blankets the land in a serene hush, inviting contemplation and introspection. Furthermore, the varying forms of snowflakes and their intricate patterns reflect the intricacies of atmospheric processes, offering insights into temperature differentials and moisture content. Finally, hail, with its sudden and intense nature, delivers a message of intensity and resilience. The formation of hailstones within thunderstorms signifies the immense energy within the atmosphere and serves as a reminder of nature's unpredictability and power. The size and density of hailstones hold clues to the dynamics of convective processes high in the sky, shedding light on the strength and volatility of atmospheric convulsions. Understanding precipitation as a language allows us to deepen our connection with the skies and glean valuable insights into the complex interplay of factors influencing weather patterns. By decoding the messages embedded within rain, snow, and hail, we gain a deeper appreciation for the intricate dialogue between the atmosphere and the earth below. Moreover, by recognizing the profound significance of these natural phenomena, we can elevate our ecological awareness and foster a harmonious relationship with the environment.

Atmospheric Phenomena and Their Meanings

As we gaze upward into the boundless expanse of the atmosphere, we encounter a myriad of mesmerizing phenomena that not only stir our senses but also convey profound messages about the state of the environment. From awe-inspiring sunsets to awe-inducing lightning storms, each atmospheric occurrence holds valuable insights waiting to be deciphered. Meteorologists and environmental researchers have long been intrigued by the intricate meanings concealed within these phenomena, recognizing their significance in comprehending weather patterns and climatic shifts.

Atmospheric phenomena encompass a broad spectrum of natural occurrences, including halo formations, mirages, and various optical illusions caused by light refraction and reflection. These enigmatic phenomena serve as dynamic indicators of atmospheric conditions and can provide vital clues regarding impending weather changes. For instance, the appearance of halos around the sun or moon often heralds the approach of a storm, while the sighting of mirages may indicate temperature differentials and air layering. Furthermore, these phenomena carry cultural and historical connotations, having sparked myths, legends, and folklore in various societies throughout history. Exploring the meanings embedded in these atmospheric phenomena allows us to delve deeper into our planet's intricate web of interconnected systems and develop a richer understanding of the complex language of nature. By decoding these visual manifestations, we can gain invaluable insights into the Earth's ever-evolving climate and enhance our ability to predict and adapt to environmental changes.

Interpreting Optical Phenomena: Halos, Rainbows, and Mirages

Optical phenomena in the sky have captivated and mystified humans for centuries. These awe-inspiring occurrences, including halos, rainbows, and mirages, are not only beautiful but also carry symbolic and scientific significance. Understanding these optical marvels can provide insights into atmospheric conditions and the behavior of light, contributing to a deeper comprehension of our natural world. Halos, which appear as luminous circles or arcs around the sun or moon, are produced by the refraction, reflection, and dispersion of light through ice crystals in high-altitude cirrus clouds. These ethereal displays have sparked various cultural interpretations, symbolizing purity, protection, or divine presence in different societies throughout history. Rainbows, characterized by their vibrant bands of colors, result from the dispersion, reflection, and refraction of sunlight in water droplets, typically observed after rain showers. Beyond their striking visual appeal, rainbows have been emblematic of hope, transformation, and serenity across diverse cultures. Additionally, mirages represent intriguing optical illusions caused by atmospheric refraction, altering the appearance of distant objects, or creating false images of bodies of water or oases in desert landscapes. These captivating illusions have inspired folklore and stories, often serving as metaphors for the elusive nature of truth and perception. Investigating and interpreting these optical phenomena not only enriches our understanding of meteorology and atmospheric optics but also unveils the deep-rooted connections between human culture and the natural world. By decoding the language of halos, rainbows, and mirages, we glean profound insights into the intertwined narratives of science, folklore, and human perception. As we explore the enchanting realm of optical phenomena in the skies, we embark on an intellectual journey that unveils the intricate tapestry of Earth's atmosph eric language.

Temporal Changes and Seasonal Messages

The study of temporal changes and seasonal messages in the language of the skies is a fascinating dive into the interconnectedness of nature's communications. Throughout history, humans have recognized the significance of celestial occurrences in marking the passage of time, forecasting weather patterns, and governing various natural phenomena. Observing the celestial bodies and understanding their roles in signaling seasonal changes has been integral to the development of early calendars and agricultural practices. The motion of the sun, moon, and stars across the sky provides valuable cues about the changing seasons and the environmental conditions they bring. From ancient civilizations to modern scientific research, the ability to decipher the temporal changes and seasonal messages encoded in the sky has been crucial for human survival and progress. The transition of the constellations, the lengthening or shortening of daylight hours, and the positioning of celestial bodies all carry profound implications for timing rituals, predicting agricultural cycles, and guiding migratory patterns of animals. Furthermore, celestial events such as equinoxes, solstices, and eclipses have held significant cultural, religious, and agricultural importance throughout the ages. Today, advancements in technology and scientific understanding enable us to unravel the intricate relationship between temporal changes in the sky and their impact on Earth. By studying celestial phenomena, atmospheric dynamics, and global climate patterns, researchers can gain insights into long-term climate trends, extreme weather events, and their implications for various ecosystems. Understanding the seasonal messages carried by the skies allows us to adapt agricultural practices, prepare for weather-related challenges, and appreciate the harmony of natural rhythms. Moreover, this knowledge enhances our capacity to address contemporary environmental concerns and develop sustainable strategies for coexisting with the dynamic forces of nature. In essence, the language of temporal changes and seasonal messages in the skies serves as a timeless guide for navigating the intricacies of our planet's interconnected ecosystem, offering profound wisdom that continues to shape our understanding of the world around us.

Conclusion: Integrating Sky Languages with Earth Communications

Understanding the language of the skies is an endeavor that leads to a deeper comprehension of the interconnectedness between celestial messages and the terrestrial world. Throughout history, humans have sought to decipher the meanings behind various celestial occurrences, from the cyclical patterns of the stars to the ever-changing forms of cloudscapes. This pursuit has not only enriched our knowledge of meteorology and atmospheric science but has also provided cultural, spiritual, and practical insights that have shaped civilizations.

As we integrate sky languages with earth communications, it becomes evident that the celestial sphere acts as a profound communicator, offering subtle cues about the environment, climate, and natural phenomena. By discerning the intricate correlations between seasonal changes, celestial positions, and atmospheric conditions, we can attain a more comprehensive understanding of terrestrial processes and their underlying influences. The rhythmic transitions of the skies, such as the shifting constellations and the solar and lunar cycles, are deeply ingrained in many cultural traditions and have guided agricultural practices, navigation, and temporal reckonings.

Moreover, the nuances of meteorological symbols, cloud formations, and atmospheric phenomena provide invaluable insights into upcoming weather patterns, climatic shifts, and ecological fluctuations. By adeptly interpreting these signals, humanity can enhance its adaptation to natural elements, mitigate disaster risks, and optimize resource management. The integration of sky languages with earth communications fosters a harmonious coexistence with the environment, propelling endeavors toward sustainability and resilience.

Furthermore, the fusion of sky languages with earth communications offers boundless opportunities for scientific exploration and technological advancements. By harnessing the data embedded within celestial events and atmospheric dynamics, researchers can refine predictive models, elucidate climate change impacts, and unravel the mysteries of cosmic interactions. Such endeavors pave the way for innovations in astronomy, meteorology, and environmental studies, driving progress in understanding our planet's place within the cosmos.

In conclusion, the integration of sky languages with earth communications signifies a convergence of ancient wisdom, modern science, and future aspirations. It represents a holistic approach to comprehending the intricate dialogues taking place above and around us. As we continue to delve into the language of the skies, let us embrace the profound teachings it offers, fostering a profound appreciation for the celestial wonders while forging a symbiotic relationship with our earthly home.

Chapter 9

Geological Scripts: Reading the Rocks

What are Geological Scripts

Geological scripts serve as invaluable records of Earth's history, encapsulating a wealth of information within their stoic forms. As we embark on the journey of deciphering these geological scripts, it is crucial to first comprehend the fundamental aspects of rock types and formations. Rocks, often regarded as nature's historians, can unveil a plethora of stories through their composition, texture, and structural attributes. The basics of rock types and formations constitute a gateway to unlocking the secrets held within the Earth's crust. By delving into the distinct categories of rocks – igneous, sedimentary, and metamorphic – and understanding their formation processes, we gain insights into the environmental conditions, climatic variations, and the intricate geologic events that have shaped our planet over myriad ages. Moreover, exploring the different ways rocks and minerals can preserve historical and environmental data enables us to reconstruct the chronicles of ancient landscapes, catastrophic events, and evolutionary transformations. Each layer of sedimentary rocks, meticulously composed over eons, narrates tales of changing climates, fossilized life forms, and even the traces of major geological upheavals. Within the intricate folds of metamorphic rocks lie imprints of immense pressures, extreme temperatures, and the profound alterations experienced by their predecessors. Igneous rocks, born from the fiery crucible of volcanic activity, bear witness to the molten forces that once sculpted our world. Together, these rock types form an elaborate library of earthly archives, waiting to be decoded and understood. The study of rock types and formations not only unravels the mysteries of Earth's past but also serves as a powerful tool for predicting potential future scenarios, including geological hazards and environmental changes. In essence, delving deep into the basics of geological scripts is akin to embarking on an enthralling expedition through time, offering glimpses into the grand narrative of our planet's evolution.

The Basics of Rock Types and Formations

Rocks are the enduring storytellers of Earth's history, each type and formation containing invaluable clues to our planet's past. One of the fundamental principles in understanding rocks is recognizing the three main rock types: igneous, sedimentary, and metamorphic. Each type bears distinct characteristics that reflect the processes by which they were formed. Igneous rocks originate from the cooling and solidification of magma or lava, exhibiting a crystalline texture. These rocks can provide insights into volcanic activity and the history of Earth's crust. Sedimentary rocks, on the other hand, are created through the accumulation and cementation of mineral and organic particles, laying down layers that capture snapshots of different eras. They often reveal evidence of ancient environments, such as ocean beds, deserts, or river deltas. Finally, metamorphic rocks form under intense heat and pressure, transforming existing rocks into new configurations while leaving behind traces of their original mineral content. Understanding these fundamental rock types allows geologists to piece together the intricate tale of our planet's development. Furthermore, the various formations within these rock types hold important clues about the conditions and forces at play during their creation. For instance, stratifications and layering in sedimentary rocks provide records of shifting environmental conditions and the passage of time. Meanwhile, the textures and mineral alignments in metamorphic rocks offer glimpses into the immense pressures and temperatures they have endured. Moreover, the presence of minerals like quartz, feldspar, mica, and calcite can provide crucial information about the processes that led to the formation of igneous rocks. By unlocking the language of rocks, geologists gain a deeper understanding of Earth's geological history, enabling them to reconstruct ancient landscapes, identify past climate patterns, and infer the dynamic processes that have shaped the planet.

Methods for Decoding Rock Formations

Decoding rock formations is a nuanced and intricate process that draws upon a diverse array of scientific disciplines. Geologists employ an assortment of methods to unravel the stories hidden within the Earth's rocky archives, seeking to understand the events and conditions that have shaped our planet over millions of years. One fundamental approach involves petrology, the study of the composition, texture, and formation of rocks. By scrutinizing the mineralogical makeup and structural characteristics of rocks, geologists can classify them into different categories and infer the processes that led to their formation. Additionally, geochemistry plays a pivotal role in decoding rock formations, as it explores the distribution and interactions of chemical elements within rocks. Through techniques such as X-ray fluorescence and mass spectrometry, geoscientists can ascertain the elemental composition of rocks, shedding light on their origins and the environmental conditions at the time of their formation. Another essential method is stratigraphy, which involves the analysis of rock layers, or strata, to comprehend the sequence and relative ages of geological events. This approach allows researchers to construct a timeline of Earth's history, discerning periods of stability, upheaval, and the evolution of life forms. In conjunction with these methods, geochronology aids in determining the absolute ages of rocks through radiometric dating techniques, offering crucial insights into the temporal aspects of geological events. Furthermore, tectonic geomorphology provides valuable information by examining the surface features resulting from tectonic forces, contributing to the understanding of how rock formations have been influenced by movements and deformations of the Earth's crust. Importantly, remote sensing technologies like LiDAR and satellite imaging facilitate the identification and mapping of rock formations over expansive areas, enriching our knowledge of geological structures and landforms. Ultimately, the integration of these multifaceted approaches enables geologists to decipher the narratives concealed within rock formations, providing profound insights into the Earth's dynamic history and the intricate processes governing its evoluti on.

Case Studies: Historical Insights from Rocks

Rock formations hold within them the secrets of Earth's ancient past, providing a window into the historical events and environmental conditions that have shaped our planet. By examining and analyzing specific rock formations, geologists can uncover invaluable insights into the geological, climatological, and biological history of the Earth. Through case studies of notable rock formations, we can gain a deeper understanding of the profound impact that these geological records have on our understanding of the planet's history and its future. One such case study is the Grand Canyon in the United States, where the exposed layers of sedimentary rocks reveal millions of years of geological history. By studying the distinct layers of rock and the fossils embedded within them, scientists have been able to reconstruct the environmental conditions and the evolution of life during different periods of the Earth's history. Another compelling case study is the Burgess Shale in Canada, which has yielded an exceptionally preserved record of marine life from the Cambrian period over 500 million years ago. The intricate details contained within the shale have provided unprecedented insights into early animal diversity and ecosystem dynamics. Furthermore, the study of volcanic rocks from different geological eras has allowed researchers to understand past volcanic activity and its impact on the environment and climate. By examining the chemical composition and mineralogical characteristics of these rocks, scientists can piece together a comprehensive picture of ancient volcanic events and their global repercussions. These case studies demonstrate the power of geological analysis in unlocking the mysteries of Earth's distant past and offer valuable lessons for interpreting present-day geological phenomena. By delving into the stories told by rocks, we can gain a deeper appreciation for the interconnectedness of Earth's history and the dynamic processes that continue to shape our world.

Sedimentary Stories: Time Encapsulated in Layers

Sedimentary rocks hold within them an awe-inspiring chronicle of Earth's history, each layer serving as a time capsule that encapsulates the events and environments of the distant past. The stratigraphic sequence within sedimentary rocks offers a unique insight into the changing landscapes, climates, and ecosystems that have shaped our planet over millions of years. By effectively unraveling these sedimentary stories, geologists can piece together a narrative that spans epochs, revealing the relentless ebb and flow of geological forces and natural phenomena.

As we examine the intricate layers of sedimentary rocks, we are presented with an opportunity to decipher the subtle nuances embedded within each stratum. These layers not only reflect the processes of deposition and lithification but also capture the essence of bygone eras. Whether it is the remnants of ancient marine life preserved in limestone or the tell-tale signs of prehistoric river systems immortalized in sandstone, each sedimentary formation carries with it a profound tale of Earth's evolution.

Moreover, the study of sedimentary rocks enables us to reconstruct paleoenvironments, granting us a glimpse into the environmental conditions that prevailed during different periods of geological history. By analyzing the composition and characteristics of these layered archives, scientists can draw conclusions about ancient climates, sea levels, and ecological dynamics, fostering a deeper understanding of the interconnected web of life throughout the ages.

Furthermore, the examination of sedimentary sequences provides invaluable insights into major geological events such as mass extinctions, tectonic upheavals, and climatic shifts. Through meticulous analysis of sedimentary records, researchers can trace the impact of catastrophic occurrences and transformative processes, unveiling the intricate tapestry of Earth's tumultuous past.

In essence, the study of sedimentary rocks serves as an indispensable tool for unraveling the mysteries of our planet's history. Within their stratified layers lie the echoes of forgotten worlds, allowing us to peer through the veil of time and gain a profound appreciation for the enduring legacy of geological processes. Through meticulous observation and interpretation, these sedimentary stories not only enrich our understanding of Earth's past but also provide valuable insights that resonate with the challenges and transformations shaping our world today.

Metamorphic Messages: Tracing Transformative Events

Metamorphic rocks hold within them the enigmatic narratives of transformative events that have unfolded beneath the Earth's surface. These rocks, formed under intense heat and pressure, carry profound stories of change and metamorphosis. Through meticulous observation and analysis, geologists can decipher the rich messages preserved in these rocks, unraveling the history of geological transformations. Metamorphic rocks often originate from pre-existing sedimentary or igneous rocks, undergoing a profound journey of alteration in response to tectonic forces and deep-seated processes. The study of these rocks provides crucial insights into the dynamic nature of the Earth's crust and the forces that shape it. By scrutinizing the textures, mineral composition, and structural features of metamorphic rocks, geologists can discern the conditions under which these enigmatic formations came into being. Consequently, they gain unprecedented access to the ancient chapters of Earth's geological saga, piecing together a narrative of immense geological significance. Furthermore, the identification of index minerals and the assessment of mineral assemblages within metamorphic rocks allow geologists to infer the temperature and pressure conditions under which these rocks were shaped, shedding light on the sheer magnitude of the geological events that took place eons ago. The process of metamorphism also often leads to the development of unique structures, such as foliation, which holds invaluable clues about the dynamic forces acting deep within the Earth. Whether it's the graceful alignment of minerals or the intricate patterns of deformation, these structures offer profound insights into the cataclysmic forces that have sculpted the Earth over millennia. Moreover, the study of metamorphic rocks carries extensive implications for understanding critical phenomena such as mountain building, plate tectonics, and the evolution of geological landscapes. Through these enduring narratives captured within metamorphic rocks, geologists are empowered to reconstruct the intricate tapestry of Earth's geological evolution, unveiling the dramatic events that have shaped our planet and continue to influence its transformation to this day.

Igneous Indications: Clues from the Magma

The study of igneous indications is a critical aspect of geological analysis, offering valuable insights into the Earth's history and processes. Igneous rocks form from the solidification of magma or lava, and their properties provide essential clues about the conditions and events that shaped them. By examining these rocks, geologists can uncover a wealth of information about past volcanic activity, tectonic movements, and the evolution of Earth's crust.

One of the key aspects of studying igneous rocks is the understanding of their mineral composition and texture. This can reveal crucial details about the cooling rate and environment in which the rock formed. For example, rocks with a fine-grained texture may indicate rapid cooling, while those with larger crystals suggest slower cooling and deeper underground formation. By analyzing the mineral content, geologists can also infer the temperature, pressure, and chemical composition of the magma that gave rise to the rock.

Furthermore, igneous rocks often display distinctive features such as vesicles, which are cavities left by gas bubbles during the cooling process. These vesicles can provide evidence of the volatile components present in the magma, shedding light on the magmatic processes and potential volcanic eruptions. In addition, the presence of certain minerals or crystal structures can indicate the prevailing conditions during the rock's formation, including the degree of partial melting, the intensity of volcanic activity, and the influence of external factors like water content and pressure.

Moreover, the study of igneous indications offers valuable insights into the Earth's past climatic conditions. For instance, ash layers and volcanic deposits within igneous rocks provide tangible records of ancient volcanic eruptions and associated environmental impacts. These deposits can help reconstruct past climates, assess the effects of volcanic events on ecosystems, and contribute to our understanding of long-term climate change. Furthermore, integrating geochemical data from igneous rocks with climatological evidence enables geologists to reconstruct historical atmospheric compositions and infer the influence of volcanic activity on global temperatures.

In summary, the analysis of igneous indications plays a pivotal role in deciphering Earth's history and unraveling the complex interactions between geological processes and environmental changes. Through careful examination and interpretation of igneous rocks, scientists can piece together a detailed narrative of the planet's turbulent past, offering invaluable perspectives on its future trajectory.

Integrating Climatological Data with Geological Findings

As we delve into the intricate web of Earth's history, the integration of climatological data with geological findings emerges as a pivotal bridge between the past and the present. In deciphering the ancient conversations etched within rocks, the role of climate dynamics cannot be understated. The patterns of sedimentation, the formation of distinct layers, and the metamorphic transformations all bear the indelible imprint of climatic influences. By aligning these geological narratives with climatological data, profound insights can be gleaned regarding the ever-changing environmental conditions that have shaped our planet over eons. From the analysis of fossilized fauna and flora to the isotopic composition of minerals, the synergistic interplay of geological and climatological evidence offers a comprehensive chronicle of Earth's bygone eras. Furthermore, the correlation of glacial movements, sea level fluctuations, and atmospheric compositions with geological archives unveils a compelling saga of natural forces at play. Through this collaborative analysis, the enigmatic puzzle of Earth's climatic history is gradually unravelled, shedding light on the intricate dance between geological processes and climatic oscillations. This interdisciplinary approach not only enriches our understanding of ancient environments but also holds profound implications for comprehending present-day climate shifts and predicting future trends. By fusing the fine-grained details of geological formations with the broader canvas of climatological patterns, researchers can construct a more holistic narrative of Earth's climatic evolution. This amalgamation of disciplines transcends the confines of temporal boundaries, fostering a deeper appreciation of Earth's dynamic equilibrium. Ultimately, the harmonious fusion of geological records with climatological datasets serves as an invaluable compass, guiding us through the labyrinthine corridors of Earth's climatic history and illuminating the path forward in our quest to comprehend the intricate tapestry of our ever-changing planet.

Implications for Understanding Mountain Formations

Mountains, as majestic geological features, provide profound insights into Earth's history and the complex interactions between tectonic forces and climatic processes. Understanding the formation of mountains holds significant implications for various scientific fields and our comprehension of the planet's dynamic nature. This section delves into the multifaceted implications stemming from the study of mountain formations.

One key implication lies within the realm of plate tectonics. By deciphering the processes that lead to the creation of mountain ranges, geologists can gain invaluable knowledge about the movement and interaction of tectonic plates. This understanding is instrumental in assessing seismic hazards and developing strategies for mitigating the impact of geological events on human populations and infrastructure. Furthermore, exploring mountain formations contributes to ongoing research into the long-term evolution of Earth's lithosphere.

Moreover, the study of mountain formations offers crucial insights into paleoclimate and environmental changes. Examining the composition and structure of mountains allows scientists to reconstruct past climatic conditions, ultimately enhancing our understanding of historical climate patterns and their influence on evolutionary processes and biodiversity. This knowledge not only enriches our grasp of Earth's ecological history but also provides valuable context for contemporary climate change studies and the potential impacts on natural ecosystems and human societies.

Additionally, mountain formations serve as archives of Earth's deep geological history. Through the analysis of mountain structures and the rocks they comprise, researchers can unravel the intricate narrative of our planet's early development and the myriad geological events that have shaped its present state. Such insights shed light on the formation of continents, the emergence of life forms, and the ebb and flow of ancient oceans, enriching our knowledge of Earth's geological chronicle.

The implications extend beyond terrestrial realms, encompassing planetary science and cosmic evolution. Mountains offer analogs for extraterrestrial landforms, aiding in the interpretation of geological features on other celestial bodies. The lessons gleaned from mountain formations enable researchers to draw parallels with phenomena observed on Mars, the Moon, and beyond, facilitating the comparative analysis of planetary geology and the investigation of the broader processes governing the formation of rocky worlds.

In essence, the study of mountain formations transcends disciplinary boundaries and yields far-reaching implications for our understanding of Earth's past, present, and future. As we continue to unravel the secrets embedded in these colossal landforms, we unravel a deeper appreciation for the interconnectedness of geological, climatological, and ecological systems, fostering a holistic comprehension of our planet's intricate tapestry.

Summary and Concluding Thoughts

In summarizing the intricate relationship between geological scripts and the formation of mountains, it becomes evident that rock formations hold a wealth of information regarding the earth's history. The study of these formations provides us with invaluable insights into the processes that have shaped our planet over millennia. As we have explored the various rock types and their significance in deciphering the earth's narrative, it is important to emphasize the remarkable role of geology in deepening our comprehension of natural phenomena. By scrutinizing sedimentary layers, we unravel a chronicle of environmental shifts and ancient landscapes, allowing us to comprehend the evolution of the earth's surface. Additionally, delving into the metamorphic nature of rocks elucidates the extensive transformational occurrences that have left enduring imprints. Through the analysis of igneous rocks, we gain an understanding of volcanic activities and the molten forces that have sculpted terrains. This comprehensive knowledge not only enriches our understanding of the past but also has profound implications for predicting future geological events and the impacts of climate change. Furthermore, integrating climatological data with geological findings offers a holistic perspective on the earth's dynamic systems and their interconnectedness. As we contemplate the significance of our findings, it becomes evident that decoding geological scripts is indispensable in comprehending the mechanisms governing mountain formations. The profound implications of this knowledge extend beyond scholarly circles as it plays a pivotal role in land management, disaster preparedness, and sustainable development. In conclusion, the culmination of our exploration underscores the pivotal role of geological scripts in understanding the tumultuous history and the ongoing evolution of the earth's majestic mountain ranges. It is a reminder of the intricate interplay between geological processes and environmental forces, guiding us toward a deeper appreciation of the earth's profound narratives.

Chapter 10

The Language of the Mountains

Mountain Linguistics

Mountain linguistics is a fascinating field of study that delves into the intricate language of geological formations. The fundamental theories and concepts behind mountain linguistics encompass an exploration of the diverse factors responsible for shaping the majestic peaks that define our landscapes. From tectonic movements to volcanic activities, the geological foundations of mountains serve as the essential vocabulary in this profound narrative. Understanding these foundations provides invaluable insights into the intricate expressions conveyed by mountains. In this section, we will embark on a journey through time, unraveling the compelling discourse embedded within mountain ranges. Through a multidisciplinary approach, we will decode the primordial dialect inscribed in the very bedrock of our planet. By examining the dynamic processes that contribute to the formation of mountains, we gain a deeper appreciation for the enduring dialogue they engage in with the Earth's forces. This interdisciplinary investigation leads us to acknowledge the interplay between geological events, climatic variations, and biological interactions, all of which coalesce to shape and elucidate the unique lexicon of mountainous terrains. As we traverse the captivating terrain of mountain linguistics, we encounter the dialects of rock formations, fault lines, and volcanic contours, each articulating a distinct tale of the Earth's evolution. Furthermore, the stratigraphic records imprinted within mountains chart the historical narratives that unfold over millions of years, reflecting the gestural languages of erosion, deposition, and metamorphism. These tales are not limited to the physical structure alone, as the biodiversity harbored within mountain ecosystems also contributes to the narrative, speaking volumes about adaptation, speciation, and ecological interconnectedness at varying altitudinal zones. Indigenous knowledge systems offer additional perspectives, underscoring the cultural significance of mountains and the folklore woven around their enigmatic conversations. By recognizing and appreciating the indigenous linguistic contributions, we gain a holistic understanding of the deep-rooted symbiosis between humans and mountains. Moreover, advancements in technological tools and methodologies have augmented our ability to decipher and interpret the dialectal nuances embedded within mountainous landscapes. Cutting-edge techniques such as LiDAR imaging, satellite mapping, and 3D modeling allow us to scrutinize and comprehend the complex vocabulary of mountain formations with unprecedented precision. This synthesis of ancient narratives and modern scientific approaches paves the way for novel insights and discoveries within the realm of mountain linguistics, fostering an immersive experience in unlocking the profound messages etched within the Earth's lofty edifices.

The Geological Foundations of Mountains

Mountains, the grand monuments of the Earth's terrain, owe their existence to a complex interplay of geological forces that have shaped the landscape over millions of years. The formation of mountains is intrinsically linked to the movement of tectonic plates, where immense pressures and forces collide, giving rise to majestic mountain ranges such as the Himalayas, Andes, and Rockies. This monumental process unfolds through tectonic plate subduction, collision, and uplift, leading to the emergence of diverse and awe-inspiring mountainous terrains across the globe.

Within this geological tapestry, mountains are captivated in a dynamic dance of creation and destruction, molded by the relentless forces of erosion, weathering, and volcanic activity. The elaborate stratigraphy of mountains reveals the intricate history inscribed in their rocky layers, each stratum bearing witness to the unfolding narrative of Earth's geological past. From the towering peaks to the deep-seated roots, mountains stand as silent custodians of the Earth's geological heritage.

Delving into the heart of mountain geology unveils a world of mineralogical diversity and structural complexity. Crystalline rocks, such as granite and gneiss, constitute a substantial part of mountain structures, embodying the enduring strength and resilience that define these natural behemoths. Moreover, the presence of metamorphic rocks, forged under intense heat and pressure, adds an additional layer of complexity to the geological fabric of mountains.

Pivotal to the understanding of mountain formation is the concept of orogeny, the geological process through which mountain ranges are created. Orogenic events signify the culmination of immense geological forces, culminating in the uplifting of colossal landmasses and the sculpting of breathtaking landscapes. Through the intricate language of geological formations, mountains communicate the enduring legacy of planetary evolution and the profound impact of geological processes on the Earth's topography.

As we unravel the geological foundations of mountains, we gain invaluable insights into the vast time scales over which these monumental formations have evolved. Each mountain range embodies a profound narrative of Earth's geological history, offering glimpses into the primordial forces that have shaped our planet. Understanding the geological language of mountains allows us to decipher the epic saga of the Earth's tumultuous past, providing a gateway to comprehend the awe-inspiring dynamism of our planet's geological heritage.

Erosion and Sedimentation: Language Written in Stone

Erosion and sedimentation are fundamental processes that shape the language of mountains over vast expanses of time. The forces of weathering, gravity, and water continually interact with geological formations, crafting intricate narratives in the very bedrock of these majestic landforms.

Erosion, driven by wind and water, gradually sculpts the features of mountains, chiseling their surfaces into breathtaking ridges, valleys, and peaks. These physical transformations are not merely aesthetically appealing; they also serve as tangible testimonies to the ongoing dialogue between the mountain and its environment. Furthermore, the resulting sediments carry within them a chronicle of the mountain's history, capturing the essence of its evolution through the ages.

Sedimentation, on the other hand, involves the deposition of eroded material, which accumulates layer upon layer, like pages in a voluminous tome. These sedimentary deposits record a wealth of information about environmental conditions, biological communities, and geological events that have left indelible imprints on the mountain landscape. By deciphering this stratigraphic archive, scientists can unravel the complex story of the mountain's past and gain insights into the forces that have shaped its present form.

In addition to preserving the history of mountains, erosion and sedimentation also play a crucial role in influencing their present-day ecological dynamics. As land is sculpted and reshaped, niches for diverse plant and animal communities are created, fostering a rich tapestry of biodiversity. The interplay of erosion and sedimentation thus perpetuates a dynamic conversation between the physical and biological elements of mountain ecosystems, underscoring the profound interconnectedness of living and non-living components in the language of mountains.

Moreover, the processes of erosion and sedimentation exhibit a delicate balance, susceptible to disruption by anthropogenic activities. Human interventions such as deforestation, mining, and construction can accelerate erosion rates and alter natural sedimentation patterns, leading to detrimental consequences for mountain ecosystems. Understanding the repercussions of human impact on these foundational geological processes is essential for the preservation and sustainable management of mountain environments.

As we contemplate the intricate lexicon of mountains, it becomes evident that erosion and sedimentation are not mere passive actors in their narrative; rather, they are active agents that dynamically shape the evolving syntax of these awe-inspiring landscapes. By delving into the language written in stone, we gain a profound appreciation for the enduring dialogues that unfold within the heart of the earth.

Mountain Flora and Fauna: Contributors to the Narrative

The intricate tapestry of mountain ecosystems is interwoven with a rich diversity of flora and fauna, each playing essential roles in the complex narrative of these elevated landscapes. From the resilient alpine plants that cling tenaciously to rocky outcrops to the elusive snow leopards that prowl the high peaks, every organism contributes its unique voice to the mountain language.

Flora, both delicate and robust, decorates the mountainsides with a vibrant display of colors and textures. The adaptation of these plants to the harsh mountain environment is an ongoing saga of resilience and resourcefulness. From the hardy alpine flowers that bloom for a fleeting moment in the brief summer to the ancient bristlecone pines that withstand centuries of harsh conditions, each holds a chapter in the story of survival at altitude.

Furthermore, the fauna of the mountains adds further depth to this narrative. The majestic snow leopard, symbol of wilderness and agility, silently stalks its prey on the steep slopes, epitomizing the grace and power demanded by life in this rugged terrain. Meanwhile, the golden eagle soars with elegant authority through the expansive skies above, casting its keen eyes on the unfolding drama below. Smaller creatures such as the pika and marmot are indefatigable in their activities, carving out a living amongst the scree and boulders, each contributing their own verse to the mountain's ballad.

The interactions between the flora and fauna in these environments create a symphony of life, each note harmonizing with those around it to form an ecosystemic melody. Pollinators flit from bloom to bloom, ensuring the continuation of the floral chorus, while herbivores navigate the landscape, sculpting the composition of the plant community through their selective grazing. Predators maintain the delicate balance, preventing any one element from overshadowing the rest. This complex web of biodiversity intertwines into a narrative fabric that reveals the interconnectedness and resilience of life in the mountains.

Understanding the various dimensions of mountain flora and fauna is therefore pivotal in deciphering the language of these landscapes. Their presence, absence, and adaptations are integral indicators of the state of a mountain ecosystem, reflecting the impacts of climate change, human influence, and other environmental factors. As we delve deeper into the language of the mountains, we must heed the insights offered by the diverse voices of its flora and fauna, recognizing them as invaluable contributors to the ongoing narrative of these awe-inspiring landscapes.

Altitudinal Zonation: Vertical Language Layers

Altitudinal zonation, or the vertical arrangement of ecological communities on a mountain, is akin to the chapters in a book, each layer telling its own story within the larger narrative of the mountain itself. As one ascends a mountain slope, dramatic changes unfold in vegetation, climate, and wildlife, each revealing a unique aspect of the mountain's language. The subtleties of these altitudinal bands hold profound insights into the complex interactions between biotic and abiotic elements. At the foothills, a rich tapestry of life unfolds, characterized by forests teeming with an array of trees, shrubs, and diverse wildlife. As altitude increases, this verdant expanse transforms into montane forests, where hardy coniferous trees reign supreme, offering refuge to elusive species adapted to the cooler climate. Reaching higher elevations, alpine tundra sets the stage, with its austere beauty and resilience in the face of harsh conditions. Here, low temperatures and relentless winds sculpt a landscape inhabited by specialized plants and animals that have evolved to thrive in extreme environments. Finally, at the lofty peaks, the barren yet mesmerizing terrain of the nival zone prevails, where only the hardiest organisms eke out an existence amidst severe cold and scant resources. The altitudinal zonation of mountains mirrors a hierarchical social structure, with each tier playing a vital role in the overall coherence of the mountain ecosystem. This stratification represents the culmination of centuries of adaptation and coevolution, manifested in a symphony of life forms, each contributing a verse to the majestic ballad of the mountain. However, human activities have begun to disrupt this harmonious arrangement, exerting pressures that reverberate through the layers of altitudinal zonation. These disturbances threaten to erode the delicate balance of these ecosystems, potentially silencing the intricate voices of the mountains. Nevertheless, technological advancements and interdisciplinary research offer hope in deciphering and preserving the nuances of altitudinal zonation. By recognizing the significance of each layer and striving to understand their distinct dialects, we can cultivate a deeper appreciation for the mountains' extraordinary array of languages, and work towards ensuring their preservation for generations to come.

Indigenous Knowledge and Mountain Speech

Indigenous knowledge plays a vital role in understanding the intricate linguistic tapestry of mountainous regions. For generations, indigenous communities residing in these areas have honed an intimate understanding of the land, its rhythms, and its messages. Their traditional ecological knowledge encompasses a rich repository of insights into the language of the mountains. Through their deep connection with nature, indigenous peoples have deciphered the subtle cues embedded in the landscape, recognizing the nuanced expressions of the land and its inhabitants. These insights often encompass an array of domains including the behavioral patterns of wildlife, the seasonal shifts, and the ecological balance - all of which contribute to the narrative of the mountains. It is through this profound awareness that indigenous communities have developed a holistic understanding of the natural world that goes beyond scientific explanations. As custodians of ancient wisdom, these communities offer invaluable perspectives on the dialects spoken by the mountains.

Their understanding is rooted in respect for the interconnectedness of life, acknowledging that every element within the mountain ecosystem contributes to a complex dialogue. Through careful observation and oral traditions passed down through the ages, indigenous groups have transmitted their knowledge - providing a roadmap to interpret the messages woven into the fabric of the mountains. Their insights also shed light on sustainable practices that have shaped their harmonious coexistence with the environment, serving as a poignant reminder of the significance of preserving traditional wisdom in a rapidly changing world.

Furthermore, acknowledging and integrating indigenous knowledge into mainstream discourse is essential for a comprehensive comprehension of mountain speech. Collaborative efforts that bridge indigenous wisdom with contemporary scientific inquiry can pave the way for a more profound insight into the profound intricacies of mountainous communication. By embracing diverse perspectives, we can foster a more inclusive dialogue that captures the multifaceted essence of mountain language, elevating our collective understanding of these majestic landscapes.

Climatic Influences on Mountainous Communication

Mountains are dynamic systems that serve as conduits for environmental communication, playing a crucial role in shaping regional and global climates. The interaction between mountains and climate is complex and multifaceted, exerting profound influences on the ecological and geological dynamics of these formidable landscapes. Climatic variables such as temperature, precipitation, wind patterns, and atmospheric pressure have a profound impact on mountainous communication, significantly shaping the diversified languages spoken by these ancient landforms. One of the most significant climatic influences on mountain communication is the phenomenon of orographic effect, which occurs when moist air is forced to rise over elevated terrain, leading to the condensation of water vapor and subsequent precipitation. This process not only results in the formation of distinct microclimates within mountain ecosystems but also contributes to the development of unique ecological communities and habitats. Additionally, the intricate interplay between mountainous topography and climatic factors gives rise to specialized weather patterns, including foehn winds and katabatic winds, further enriching the lexicon of mountainous communication. Furthermore, the seasonal variability of climatic conditions exerts a profound influence on the rhythmic oscillations of mountain speech, fostering the emergence of distinctive dialects characterized by seasonal cues and nuances. For instance, the annual thawing of snow and ice in high-altitude regions heralds the arrival of new narratives, symbolizing renewal and rebirth within the language of the mountains. Conversely, the onset of harsh winter conditions evokes tales of endurance, resilience, and adaptation within the timeless chronicles of these majestic landscapes. Moreover, the interconnectedness between mountainous terrains and global climatic patterns underscores the critical role of mountains in regulating atmospheric circulation and influencing regional weather phenomena. As agents of atmospheric moisture redistribution, mountains contribute to the modulation of precipitation regimes and hydrological cycles, modulating the flow and expression of the hydrological language inscribed across their slopes and valleys. Hence, understanding the climatic influences on mountainous communication is paramount in unraveling the intricate web of interactions that characterize these enigmatic landscapes, offering invaluable insights into the symbiotic relationship between mountains and climatic dynamics.

Human Impact on Mountain Languages

The delicate and intricate balance of mountain ecosystems has been significantly impacted by human activities, leading to profound effects on the linguistic expressions of these majestic landscapes. Human impact on mountain languages is a multifaceted issue encompassing a range of activities that have led to transformations in these ecological narratives. The sustainable coexistence of humans and mountains hinges on understanding and mitigating these impacts.

One of the most evident human impacts on mountain languages is deforestation. Irresponsible logging practices and clearing of land for agriculture or development have eroded the rich vocabulary of mountain ecosystems. As age-old forests are cleared, the nuanced language of the flora and fauna is silenced, altering the natural dialogue that once thrived among the diverse species residing in these mountains.

Furthermore, urbanization and infrastructure development in mountainous regions have accelerated the dissonance in mountain languages. Expanding human settlements, road networks, and industrial complexes disrupt the harmonious communication channels intrinsic to these landscapes. These interventions not only fragment habitats but also introduce foreign elements into the delicate discourse of mountain ecosystems, leading to irreversible changes in their linguistic patterns.

Additionally, human-induced climate change poses a formidable threat to mountain languages. Increasing temperatures, altered precipitation patterns, and glacial retreat endanger the delicate balance of mountain ecosystems, causing species displacement and extinction. These shifts in ecological dynamics reverberate in the linguistic fabric of mountains, as traditional expressions undergo modification or fade into silence.

The extraction of natural resources from mountainous terrains has further exacerbated the human impact on mountain languages. Mining operations, hydroelectric projects, and commercial exploitation of natural wealth have left scars on the mountains and altered their linguistic repertoire. The loss of mineral-rich soils, water sources, and precious biodiversity has irreversibly transformed the narrative of mountain languages, diminishing their richness and complexity.

Addressing the human impact on mountain languages necessitates a holistic approach that encompasses conservation, sustainable development, and community engagement. By fostering awareness and promoting responsible stewardship of mountain ecosystems, it is possible to mitigate these impacts and endeavor to restore the integrity and vibrancy of mountain languages. Through collaborative efforts, it is imperative to ensure that these ancient dialogues continue to resonate across the peaks and valleys, embodying the profound interconnectedness of humans and mountains.

Technological Advances in Mountain Studies

As our understanding of mountain environments deepens, technological advancements have played a pivotal role in enhancing our ability to study and decipher the complex language of the mountains. Remote sensing technologies, including satellite imagery and LiDAR (Light Detection and Ranging), have revolutionized the way we observe and analyze mountain landscapes. These tools offer unprecedented insight into the topography, geomorphology, and ecological dynamics of mountains, allowing researchers to map and monitor changes with remarkable precision. Furthermore, geographic information systems (GIS) enable the integration of diverse spatial data, facilitating comprehensive spatial analysis and modeling of mountain processes and phenomena. The utilization of drones equipped with high-resolution cameras and sensors has expanded our capacity to capture detailed imagery and collect specific data points from remote and rugged mountainous terrain. This has proven invaluable for assessing vegetation distribution, wildlife habitats, and geological features in otherwise inaccessible areas, contributing substantially to our knowledge of mountain ecosystems. In addition, advances in molecular biology and genetics have empowered scientists to uncover intricate details about the biodiversity and evolutionary adaptations of mountain species. DNA sequencing and genomic tools have shed light on the genetic diversity, population structures, and adaptive traits of flora and fauna inhabiting mountain regions, offering critical insights into their resilience and vulnerability in the face of environmental changes. Coupled with traditional fieldwork and ecological surveys, these innovative techniques have enriched our comprehension of mountain biota. Furthermore, sophisticated climate models and simulation software have enabled researchers to simulate and forecast the impacts of climate change on mountain systems, providing valuable projections for informing policy decisions and conservation strategies. The integration of big data analytics and machine learning algorithms has facilitated the processing and interpretation of large datasets derived from mountain observations, allowing for more robust and nuanced interpretations of complex interactions within mountain ecosystems. As technology continues to advance, there is immense potential for the continued refinement and expansion of methodologies for studying mountain environments, ensuring that we are equipped to preserve and sustain these unique and vital landscapes for generations to come.

Synthesis and Future Directions

As technological advancements continue to revolutionize the field of mountain studies, they open up new doors for research and exploration. The synthesis of data collected through these cutting-edge tools enables us to gain a more comprehensive understanding of the language of the mountains. With the aid of satellite imagery, LiDAR technology, and GIS mapping, researchers can delve deeper into the intricate details of mountainous landscapes, uncovering hidden patterns and anomalies that were once elusive. These technologies have not only enhanced our ability to analyze the geological structure of mountains but have also facilitated the monitoring of environmental changes in these regions. As we move forward, the integration of artificial intelligence and machine learning algorithms will further amplify our capacity to process and interpret vast amounts of data, providing valuable insights into the complex dynamics of mountain systems.

Moreover, future directions in mountain studies call for interdisciplinary collaboration and a holistic approach to research. By engaging experts from various fields such as geology, ecology, climatology, anthropology, and indigenous knowledge systems, we can construct a more interconnected narrative of mountain languages. Embracing diverse perspectives and methodologies will enrich our understanding of the nuanced interactions between geological, biological, and climatic factors in shaping mountain ecosystems and their communicative processes.

Looking ahead, it is imperative to consider the implications of climate change on mountain languages. The projected impacts of changing temperatures, glacial retreat, and altered precipitation patterns underscore the urgency of studying and preserving these unique linguistic landscapes. By forecasting potential shifts in mountain ecosystems, we can anticipate the repercussions on local communities, biodiversity, and hydrological systems, thereby informing sustainable conservation strategies and adaptive management efforts. Furthermore, understanding the cultural significance of mountains in various societies and the intricate relationships between humans and these majestic terrains will be crucial for developing ethically informed approaches to mountain research and conservation.

In conclusion, the synthesis of technological innovations, interdisciplinary cooperation, and a proactive stance on environmental challenges will propel the field of mountain studies into a new era of discovery and preservation. By unraveling the lexicon of mountain languages and advocating for their safeguarding, we can contribute to the conservation of these ancient narratives while fostering an enriched appreciation for the profound beauty and ecological significance of mountainous regions.

Chapter II

Oceans and Rivers: Conduits of Ancient Conversations

Aquatic Linguistics

The study of language in aquatic environments offers a unique and captivating lens through which to understand the intricate communications that occur beneath the water's surface. Just as human languages are shaped by complex patterns and structures, the languages of oceans and rivers reflect their own distinctive systems of communication. From the rhythmic lapping of waves against the shore to the melodious calls of marine creatures, each aspect of aquatic linguistics embodies an ancient and enigmatic conversation waiting to be decoded. The foundational concepts that guide this fascinating field of study encompass not only the sounds and gestures within aquatic ecosystems but also extend to the chemical, visual, and tactile modes of communication that play integral roles in shaping their linguistic landscapes. By delving into these fundamental principles, researchers can unveil the interconnected web of discourse that permeates the watery realms, illuminating the richness and complexity of aquatic languages. Fundamental to aquatic linguistics is the acknowledgment of the diverse range of communicative expressions present within oceanic and riverine environments, transcending traditional human-centric perceptions of language to embrace a broader spectrum of sensory, symbolic, and behavioral signals that form the essence of underwater dialogues. As explorers of aquatic linguistics venture deeper into these uncharted waters, they unravel the subtle nuances and intricacies of natural dialects, unlocking profound insights into the ecological, evolutionary, and cultural dimensions of aquatic communication. This nuanced approach equips scholars with the tools to decipher the submerged lexicons, unwinding the tales of time-honored conversations that have echoed throughout the depths of Earth's waterways since time immemorial. Combining scientific rigor with a profound sense of wonder, the study of aquatic linguistics holds the promise of reshaping our understanding of the living, breathing tapestry that unfolds beneath the surface of our planet's vast aquatic domains.

Historical Overview of Oceanic and River Communications

The historical relationship between human civilizations and aquatic environments has been paramount in shaping cultural, economic, and environmental narratives throughout the ages. From ancient seafaring communities to the river-based empires of antiquity, humanity has long recognized the vital role of oceans and rivers as mediums for communication, trade, and sustenance. This section delves into the rich tapestry of historical oceanic and river communications, exploring how these water bodies have served as conduits of knowledge, commerce, and cultural exchange. Ancient maritime cultures, such as the Phoenicians, Greeks, and Polynesians, mastered the art of ocean navigation, utilizing celestial cues and indigenous knowledge to traverse vast expanses of open water. Their maritime exploits not only facilitated trade and exploration but also fostered cross-cultural interactions, leading to the exchange of goods, technology, and ideas. Similarly, riverine civilizations like the Nile Valley, the Indus Valley, and Mesopotamia flourished due to the fertile land and waterborne trade routes provided by their respective rivers. These waterways acted as lifelines, enabling the development of advanced agricultural systems, urban centers, and interconnected societies. In addition to facilitating transportation and trade, oceans and rivers have also played a profound symbolic and spiritual role in the mythologies and belief systems of various cultures. Many ancient civilizations revered water bodies as deities or sources of divine inspiration, attributing mystical qualities to their waters and incorporating them into religious rituals and ceremonies. Whether viewed as benevolent providers of life or powerful forces to be reckoned with, the historical significance of oceans and rivers as conveyors of cultural, spiritual, and economic wealth cannot be overstated. This historical overview thus serves as a precursor to our exploration of aquatic languages, laying the foundation for understanding the enduring influence of water on human civilization and the natural world.

Decoding the Patterns: Currents and Tides

The study of oceanic currents and tides has long been an integral part of understanding the intricate language of Earth's waters. This section delves deep into the fascinating world of hydrodynamics to decipher the rhythmic dance of currents and tides that shape our planet's aquatic environment. Currents, influenced by powerful forces such as wind, temperature, salinity, and the Earth's rotation, play a crucial role in not only redistributing heat around the globe but also in shaping marine habitats and influencing weather patterns. Tides, on the other hand, are the result of the gravitational pull of celestial bodies, primarily the moon and the sun, leading to the mesmerizing ebb and flow of coastal waters. As we explore the complex mechanisms behind these phenomena, it becomes evident that currents and tides act as conduits for the circulation of nutrients and energy, nurturing diverse ecosystems and fostering essential biodiversity. Furthermore, their impact extends beyond the realm of marine biology; they significantly influence human activities such as shipping, fishing, and coastal infrastructure. By comprehending the intricate patterns of currents and tides, we gain valuable insights into the functioning of the planet's interconnected systems and the crucial role they play in sustaining life. Through cutting-edge technologies and interdisciplinary research, scientists are unraveling the language of currents and tides, enabling us to anticipate and respond to the changing dynamics of our oceans and rivers with unprecedented precision. The emerging field of predictive oceanography is revolutionizing our ability to forecast events such as storm surges, harmful algal blooms, and the movement of marine debris, thereby enhancing our capacity to mitigate potential risks and protect vulnerable coastal communities. In essence, the study of currents and tides forms a vital link in our ongoing endeavor to comprehend the language of Earth's aquatic realms, providing profound insights that extend far beyond the shor eline.

Symbolism in Water: Cultural Interpretations

Water has long been a source of inspiration and symbolism across cultures worldwide. From ancient mythology to modern literature, the significance of water as a symbol is profound and multifaceted. It represents life, purity, rebirth, and spirituality in various traditions, often embodying deeper meanings that transcend mere physical existence. The role of water as a symbol in cultural narratives reflects the diverse ways in which human societies have perceived and interacted with their natural environments. In many indigenous belief systems, water is revered as a sacred element, linking the earthly realm with the spiritual world. Its fluidity and adaptability mirror the ever-changing nature of human life and the complexities of existence. Drawing from a rich tapestry of folklore and legend, different cultures have woven elaborate stories around water, incorporating its symbolism into rituals, ceremonies, and artistic expressions. For instance, in some traditions, immersing oneself in water symbolizes spiritual purification and renewal, signifying a symbolic cleansing of the soul. Water features prominently in creation myths, often serving as a primordial element from which all life emerges, emphasizing its indispensable role as a sustainer of life and a symbol of fertility. Furthermore, the symbolism of water extends beyond its spiritual connotations to encompass themes of emotional depth and psychological introspection in literature and the arts. Writers and artists frequently utilize water as a metaphor for the subconscious mind, employing its enigmatic depths to represent the hidden recesses of human consciousness and the mysteries of the human experience. Whether depicted as calm and serene or stormy and turbulent, water, through its varied symbolism, offers a lens through which to explore the complexity of human emotions and the ebb and flow of life itself. As we delve into the cross-cultural interpretations of water symbolism, it becomes evident that its metaphorical significance transcends geographical and historical boundaries, resonating with universal themes of human existence. Reflecting on the myriad ways in which water has been imbued with meaning in diverse cultural contexts, we gain a deeper understanding of the profound relationship between humanity and the natural world.

Aquatic Life Forms and Their Roles in Ecosystem Communication

Aquatic ecosystems host a myriad of life forms, each playing an indispensable role in the intricate web of communication that defines these environments. From the microscopic phytoplankton to the majestic marine mammals, every organism contributes to the exchange of information vital for the sustenance and balance of aquatic ecosystems. Phytoplankton, through the process of photosynthesis, not only generate oxygen but also release chemical cues that dictate the behavior of other organisms in the food chain. Zooplankton, as the primary consumers, serve as crucial intermediaries in transferring energy and nutrients from phytoplankton to higher trophic levels, forming a fundamental component of the aquatic communication network. Moving up the food chain, fish species utilize a diverse array of visual cues, vocalizations, and chemical signals to navigate their environment, establish territories, locate mates, and evade predators.

The intricate courtship rituals of various fish species not only propagate their populations but also contribute to the overall harmonic communication within aquatic communities. Furthermore, marine invertebrates such as corals engage in symbiotic relationships with algae, where chemical signaling and nutrient exchange are essential components of their interaction, influencing the broader health of coral reef ecosystems. The significance of marine mammals cannot be understated, as their complex vocalizations and migratory patterns directly impact the dynamics of oceanic communication networks. From the haunting songs of humpback whales to the sophisticated echolocation abilities of dolphins and porpoises, these creatures utilize sound as a powerful tool for navigation, communication, and hunting. Additionally, sea otters play a critical role in preserving kelp forest environments by preying on sea urchins, thereby maintaining the delicate balance of this ecosystem. Overall, the myriad interactions and communications among aquatic life forms underscore the delicate interconnectedness and resilience of these ecosystems, emphasizing the necessity for conservation efforts aimed at preserving these invaluable natural resourc es.

Rivers as Historical Storytellers

Rivers, flowing through the earth's landscape for millennia, have been silent witnesses to the historical evolution of civilizations, cultures, and ecosystems. Within their meandering paths and restless currents, rivers conceal profound narratives that offer remarkable insights into the past. Each river holds within its waters a rich tapestry of history, bearing witness to the rise and fall of ancient civilizations, the migrations of peoples, and the interplay between human societies and the natural world.

As conduits of both life and civilization, rivers have served as lifelines for countless communities, shaping human settlements, trade routes, and cultural exchanges. From the majestic Nile in Africa to the iconic Ganges in South Asia, rivers have played pivotal roles in the development of human history. Their banks have hosted the birth of great empires, witnessed the spread of agriculture, and fueled the progress of technology and trade. The historic significance of rivers is immeasurable, as evidenced by the archaeological treasures unearthed along their shores, the folklore and mythology they have inspired, and the spiritual reverence they continue to command.

Moreover, rivers carry within them the sedimentary archives of bygone eras, preserving invaluable clues about past climates, landscapes, and ecological transformations. By unraveling the geological narratives embedded in riverbeds and deltas, scientists can reconstruct ancient environments and understand the forces that have shaped the earth over eons. As such, rivers stand as veritable historical libraries, offering a chronicle of the earth's transformative journey through time.

In the contemporary era, as mankind grapples with the challenges of environmental sustainability and climate change, the study of rivers as historical storytellers assumes heightened significance. Utilizing interdisciplinary approaches that combine geology, archaeology, anthropology, and hydrology, researchers can delve into the annals of history inscribed in riverine landscapes. By understanding the complex interactions between human societies and rivers across diverse temporal and spatial scales, we gain valuable perspectives on the intricate dynamics of human-nature relationships and their implications for the future.

In conclusion, to neglect the historical wisdom held by rivers would be to overlook a crucial dimension of our shared human heritage. Recognizing rivers as repositories of historical knowledge enriches our comprehension of the past and enlightens our path forward. As we continue to explore and interpret the timeless messages encoded within the flowing waters, we unveil the profound legacy of rivers as historical storytellers, fostering an enduring connection between the contemporary world and the echoes of antiquity.

Oceanic Cycles and Their Global Impact

The interconnected web of oceanic cycles plays a profound role in shaping the Earth's climate and environment, with far-reaching impacts on global ecosystems and human societies. These dynamic systems, encompassing phenomena such as El Niño, La Niña, the North Atlantic Oscillation, and the Pacific Decadal Oscillation, exert significant influence on weather patterns, sea surface temperatures, and precipitation levels across the planet. The thermal inertia and vast capacity for heat storage within the oceans contribute to moderating the world's climate, with immense implications for regional climates and weather extremes. The interplay between oceans and the atmosphere also influences atmospheric circulation patterns, affecting the distribution of heat and moisture across different regions. Furthermore, oceanic cycles play a pivotal role in regulating marine ecosystems, influencing the distribution and abundance of species, as well as the productivity of fisheries. By understanding these intricate cycles, scientists can gain valuable insights into the complex mechanisms that underpin our planet's natural systems, providing crucial knowledge for sustainable resource management and conservation strategies. Moreover, oceanic cycles have critical implications for human societies, particularly in vulnerable coastal regions. The impact of cyclical phenomenon on sea levels, storm surges, and ocean currents directly affects coastal communities and infrastructure. Anticipating and adapting to the variability and potential changes in these oceanic cycles is essential for effective coastal planning and hazard mitigation, especially in the context of accelerating climate change and rising sea levels. Research into oceanic cycles and their global impact constitutes a vital area of study with multifaceted significance, encompassing scientific, environmental, and societal dimensions. The ongoing exploration and understanding of these complex systems offer promise for enhancing our ability to manage and respond to the challenges and opportunities presented by the intricate dynamics of the world's oceans .

Technological Advances in Underwater Research

Technological advancements have revolutionized the field of underwater research, enabling scientists to explore the depths of oceans and rivers with unprecedented precision and clarity. The development of state-of-the-art submersibles equipped with advanced robotic arms and high-resolution cameras has allowed researchers to observe and document aquatic ecosystems in more detail than ever before. In addition to submersibles, remotely operated vehicles (ROVs) have become essential tools for underwater exploration, providing real-time footage and collecting samples from remote and challenging environments. These tools not only facilitate the observation of marine life but also aid in the study of underwater geological formations, helping scientists unravel the mysteries of the ocean floor.

Moreover, the use of autonomous underwater vehicles (AUVs) has dramatically expanded our ability to survey vast areas of the ocean and conduct long-term monitoring of environmental changes. Equipped with sophisticated sensors and navigational systems, AUVs can autonomously map the seafloor, collect data on water chemistry, and track marine animal behaviors, shedding light on the interconnected web of underwater communications. Additionally, advancements in satellite technology have enabled the monitoring of ocean surface temperatures, currents, and salinity levels on a global scale, providing vital information for understanding the intricate dynamics of oceanic systems.

Furthermore, the deployment of acoustic technologies, such as hydrophones and sound wave mapping, has facilitated the study of underwater soundscapes, including the communication patterns of marine mammals and the intricate vocalizations of aquatic species. This has opened new avenues for decoding the language of marine life and understanding the acoustic signals that form the basis of aquatic conversations.

In recent years, the integration of artificial intelligence and machine learning algorithms has enhanced the analysis of vast amounts of underwater data, allowing researchers to identify complex patterns and trends within aquatic ecosystems. These technologies have enabled scientists to uncover hidden connections between environmental factors, aquatic organisms, and natural phenomena, providing invaluable insights into the ancient conversations taking place beneath the waves. As we continue to push the boundaries of underwater research, innovative technologies will play a pivotal role in deepening our understanding of the profound and age-old dialogues that unfold within the world's oceans and rivers.

Case Studies: Significant Discoveries in Aquatic Communications

Throughout history, the study of aquatic communications has yielded numerous significant discoveries that have enriched our understanding of the complex interactions taking place within oceans and rivers. These case studies offer compelling insights into the diverse forms of communication found in aquatic environments, shedding light on the profound implications for ecological balance, cultural symbolism, and technological innovation.

One landmark case study involves the acoustic communication of marine mammals, particularly the intricate vocalizations of whales. Research has unveiled the existence of distinct dialects among whale populations, demonstrating the sophisticated nature of their communication and social structures. By analyzing these vocalizations, researchers have been able to decipher unique patterns and messages, providing invaluable knowledge about the behavioral dynamics of these majestic creatures.

In another compelling case study, the exploration of bioluminescent communication among deep-sea organisms has revealed mesmerizing displays of light and color used for signaling and camouflage. This extraordinary phenomenon not only showcases the astonishing adaptability of marine life but also underscores the intricate language encoded in these radiant displays. The study of bioluminescence has far-reaching implications for fields such as security technology and medical imaging, offering inspiration for innovations rooted in the natural world.

Furthermore, an in-depth investigation into coral reef ecosystems has unraveled a captivating tapestry of chemical signaling and symbiotic relationships. The intricate dance of olfactory cues and allelopathic interactions among coral and other marine species demonstrates the rich vocabulary of chemical communication underpinning the interconnectedness of these vibrant underwater communities. These discoveries have illuminated the delicate balance of life in the oceans, emphasizing the critical need for conservation efforts and sustainable practices.

The underwater world continuously astounds with its intricate modes of communication, and these case studies represent only a fraction of the remarkable findings that have expanded our comprehension of aquatic languages. As we delve deeper into these submerged realms, each discovery serves as a testament to the boundless complexity and beauty woven into the fabric of aquatic communications, inspiring ongoing exploration and reverence for our planet's watery domains.

Future Perspectives: The Next Frontier in Water-Related Research

Exploring the future of water-related research beckons us into a realm of infinite possibilities and uncharted territories. As we stand on the precipice of technological advancement, it's imperative to recognize the potential avenues for exploration that lie ahead. The field of aquatic communications presents an array of opportunities for innovative research, with numerous areas ripe for further investigation. One such area involves delving deeper into the intricate language of marine mammals, uncovering the nuances of their communication and social structures. Understanding these complex systems could provide insights into evolutionary biology and even shed light on the development of human language. Moreover, the burgeoning field of aquatic acoustics offers immense promise, as advancements in technology enable us to capture and interpret the intricate soundscape of our oceans and rivers. By harnessing cutting-edge recording and analysis techniques, researchers can unravel the rich tapestry of acoustic signals exchanged by marine life, paving the way for a deeper understanding of underwater communication networks. Another compelling frontier lies in the realm of bioluminescence, where new methodologies hold the key to deciphering the mesmerizing light patterns displayed by countless aquatic organisms. Unraveling the significance of these displays could revolutionize our comprehension of interspecies interactions and ecological dynamics within aquatic ecosystems. Furthermore, as climate change continues to reshape our planet, there is an increasing urgency to investigate its impact on aquatic environments. By integrating advanced modeling and monitoring technologies, scientists can forecast the repercussions of environmental shifts on oceanic and riverine ecosystems, providing crucial insights for conservation efforts and sustainable resource management. The future of water-related research also stands to benefit from interdisciplinary collaborations, where fields such as bioinformatics, artificial intelligence, and materials science converge to unlock new frontiers. By leveraging this synergy, researchers have the potential to develop sophisticated tools for deciphering aquatic languages, mapping intricate ecological networks, and engineering novel solutions inspired by nature. These next-generation approaches hold the promise of transforming our understanding of aquatic communication and the myriad interconnected processes that shape our planet's waterways. As we embark on this journey towards unraveling the mysteries of aquatic communications, it is imperative to embrace a multidisciplinary mindset and foster a spirit of scientific curiosity and innovation. The future vistas of water-related research are brimming with transformative potential, offering a tapestry of exploration that promises to enrich our understanding of the profound conversations unfolding beneath the waves.

Chapter 12

Harnessing the Power of Earth's Energy

Earth's Energies

The quest for sustainable and renewable sources of energy has spurred a renaissance in harnessing the power of Earth's natural resources. As we grapple with the environmental impact of traditional energy production, the need to explore diverse and sustainable energy options has become paramount. In this chapter, we embark on an exploration of the myriad ways in which the Earth offers us abundant and renewable energy resources. From the scalding depths beneath the Earth's crust to the gentle sway of wind turbines dotting the landscape, we will delve into the incredible diversity of energy sources that our planet provides. Understanding and leveraging the potential of these resources is crucial as we strive to build a more sustainable and environmentally conscious future. Through this comprehensive exploration, readers will gain insight into the pivotal role that Earth's energies play in shaping our energy landscape and the broader context of global sustainability efforts. By examining the untapped potential of geothermal, solar, wind, hydro, bioenergy, tidal, and wave power, we aim to underscore the critical importance of transitioning to renewable energy sources. This section seeks to illuminate not only the technical aspects of harnessing Earth's energies but also the compelling narrative of hope and resilience that emerges when tapping into the Earth's natural abundance. As such, this journey through the diverse types of energy derived from the Earth serves as a testament to the transformative power of renewable resources in addressing the pressing energy challenges of our time.

Fundamentals of Geothermal Energy

Geothermal energy, a reliable and sustainable source of power, is derived from the Earth's heat stored beneath its surface. The utilization of geothermal resources offers a promising alternative to traditional fossil fuels, contributing to the diversification and decentralization of energy sources.

The primary principles underlying geothermal energy revolve around the natural heat emanating from the Earth's core. Deep within the planet, radioactive decay of minerals generates intense heat, which warms adjacent rocks and water reservoirs. This heated water gives rise to geothermal reservoirs, forming the basis for geothermal power generation.

Commercially, geothermal energy is harnessed through various methods, such as hydrothermal convection systems and enhanced geothermal systems (EGS). Hydrothermal plants utilize naturally occurring geothermal reservoirs where hot water or steam can be directly extracted to drive turbines and generate electricity. In contrast, EGS involve creating deeper reservoirs by fracturing impermeable rock formations and then injecting water to extract the produced heat. This process significantly expands the geographical reach of geothermal power, enabling exploitation in areas previously considered unsuitable for conventional geothermal applications.

Furthermore, the environmental benefits of geothermal energy are substantial. This renewable resource produces minimal greenhouse gas emissions compared to fossil fuel-based power generation. It also has a smaller ecological footprint, with relatively little impact on land use.

Overcoming technical and logistical challenges remains integral to maximizing the potential of geothermal energy. Effective exploration and drilling techniques are crucial for identifying ideal locations and tapping into geothermal reservoirs efficiently. Furthermore, optimizing heat extraction and conversion processes, such as improving the efficiency of geothermal power plants, is essential for enhancing the overall economic viability of this energy source.

The future prospects for geothermal energy are bright, with ongoing research focused on advancing technology and reducing costs associated with geothermal power generation. As we continue to navigate the transition towards sustainable energy sources, the role of geothermal energy will undoubtedly become increasingly prominent, offering a valuable contribution to the global energy landscape.

Harnessing Solar Power: Techniques and Technologies

Solar power, a key player in the renewable energy landscape, offers immense potential for sustainable energy production. The harnessing of solar energy involves diverse techniques and technologies that are continuously evolving to maximize efficiency and accessibility. Photovoltaic (PV) technology stands as a frontrunner in the domain of solar power generation. By utilizing semiconducting materials, PV cells convert sunlight directly into electricity through the photovoltaic effect. This method is frequently implemented in solar panels, enabling the direct capture and conversion of solar energy. Additionally, concentrating solar power (CSP) systems harness solar radiation by using mirrors or lenses to concentrate a large area of sunlight into a small beam, which in turn produces heat. This heat is then used to generate electricity through conventional steam cycles or other processes. The advancement of CSP technology has led to the development of innovative storage solutions, such as molten salt thermal storage, ensuring a continuous and reliable energy supply even during periods of low sunlight. Moreover, emerging thin-film solar technologie s offer flexible and lightweight alternatives to traditional crystalline silicon-based solar panels, catering to various applications including building-integrated photovoltaics (BIPV) and portable solar devices. Beyond these, pioneering research is continuously exploring advanced materials and methods aimed at enhancing the performance and cost-effectiveness of solar cells. Furthermore, smart grid integration and energy management systems play a crucial role in optimizing the utilization of solar power, facilitating seamless distribution, consumption, and storage. These advancements not only enhance the reliability and resilience of solar energy systems but also contribute to the overall sustainability and stability of the energy grid. As the solar power sector continues to expand, ongoing research and technological innovation remain pivotal in unlocking the full potential of this abundant and clean energy source, driving the global transition towards a more sustainable and harmonious energy landscape.

Wind Energy: Capturing the Atmosphere's Might

Wind energy, a formidable force of nature, has been harnessed by humankind for centuries. Through innovative advancements in technology and engineering, the harnessing of wind power has become an integral component of sustainable energy production. Capturing the kinetic energy present in the movement of air masses, wind turbines have emerged as a reliable and eco-friendly source of electricity generation. The utilization of wind energy involves the transformation of wind power into mechanical or electrical energy through the use of wind turbines. These turbines are strategically positioned in areas characterized by consistent and substantial wind flow. The design and construction of these turbines necessitate a deep understanding of aerodynamics and materials science to ensure optimal performance and longevity. As the turbine blades rotate, they convert the kinetic energy of the wind into mechanical energy, which is subsequently transformed into electrical energy through a generator. Advantages of wind energy include its renewable nature, abundant availability, and minimal environmental impact compared to conventional forms of energy production. Additionally, wind energy can contribute significantly to reducing greenhouse gas emissions, thus mitigating the adverse effects of climate change. While wind energy presents myriad benefits, challenges such as intermittency and variability of wind patterns must be addressed through advanced energy storage and grid integration solutions. As technology continues to evolve, research and development efforts are driving innovations in wind energy, resulting in more efficient turbines and increased cost-effectiveness. The future of wind energy holds great promise, with ongoing initiatives focused on optimizing wind farm design, enhancing blade efficiency, and streamlining manufacturing processes. Bolstered by government incentives and public support for clean energy initiatives, the proliferation of wind power continues to gain momentum globally, positioning it as a vital component of the transition towards a sustainable energy landscape.

Hydropower: Utilizing Water for Energy Production

Hydropower, also known as water power, is a pivotal source of renewable energy that has been harnessed for centuries to generate electricity. The principle behind hydropower is relatively simple yet highly effective: the kinetic energy of flowing or falling water is converted into mechanical or electrical energy. This process involves the utilization of the natural gravitational force of water to turn turbines connected to generators, producing electricity in the process.

The history of hydropower dates back to ancient times when water wheels were used to grind grain and perform various other tasks. Today, however, hydropower plays an essential role in modern electricity generation, providing a clean and sustainable alternative to fossil fuel-based power plants. The development of hydroelectric technologies has diversified, enabling the utilization of river flows, dammed reservoirs, and even ocean tides to drive turbines and generate electricity.

One of the key advantages of hydropower is its reliability and consistency. Unlike solar or wind power, which are intermittent energy sources dependent on weather conditions, hydropower can provide a steady and predictable supply of electricity. Furthermore, it produces minimal greenhouse gas emissions, making it an environmentally friendly energy option. The construction of large-scale hydropower projects may require significant investment and careful environmental impact assessments, but the long-term benefits in terms of clean energy production and reduced reliance on non-renewable resources make it a compelling choice for many nations.

Hydropower plants can be classified into two main types: storage and run-of-river systems. Storage systems involve the construction of dams to impound water in reservoirs, regulating the flow to optimize energy production. Run-of-river systems, on the other hand, do not require large-scale impoundments and rely on the natural flow of rivers and streams to generate electricity. Both approaches have distinct advantages and considerations, and the selection of the appropriate system depends on factors such as geography, cost, and environmental impacts.

In addition to its direct utility in electricity generation, hydropower also provides ancillary benefits such as flood control, irrigation, and the creation of recreational opportunities through the formation of reservoirs. As global efforts to transition to sustainable energy intensify, the role of hydropower in the clean energy landscape continues to gain prominence. Investments in research and technological innovations are further enhancing the efficiency and environmental compatibility of hydropower, paving the way for a future where water becomes an even more vital source of renewable energy.

Bioenergy from Organic Materials

In the quest for sustainable and renewable energy sources, bioenergy derived from organic materials has emerged as a promising avenue. This section delves into the diverse mechanisms and applications of bioenergy, showcasing its potential to transform the global energy landscape. At its core, bioenergy entails the conversion of biological materials into usable energy forms, ranging from solid biomass to biogas, biofuels, and other advanced bio-based products. The process involves harnessing the energy stored within organic matter, such as agricultural residues, woody biomass, dedicated energy crops, and organic waste streams, through various technologies like anaerobic digestion, gasification, and biochemical conversion. Bioenergy's versatility is evident in its multiple applications across sectors. In the transport domain, biofuels offer an eco-friendly alternative to traditional fossil fuels, reducing greenhouse gas emissions and mitigating environmental impact. Concurrently, bioenergy plays a pivotal role in heat and power generation, fostering decentralization and resilience in energy systems. Furthermore, bioenergy contributes to circular economy models by converting organic wastes into valuable resources. With ongoing advancements in biotechnology and bio-refining, the specificity and efficiency of bioenergy processes continue to evolve. However, the widespread adoption of bioenergy faces challenges linked to resource availability, competition with food production, land use impacts, and technological complexities. Through strategic policies and interdisciplinary collaborations, these hurdles can be navigated to realize the full potential of bioenergy. As global efforts intensify to combat climate change and transition towards low-carbon economies, bioenergy stands poised to bolster energy security, rural development, and environmental sustainability. By comprehensively addressing the economic, social, and environmental dimensions, bioenergy from organic materials holds immense promise in shaping a greener and more resilient energy paradigm.

Innovative Harnessing of Tidal and Wave Power

Tidal and wave power represent an exciting frontier in renewable energy. The ebb and flow of ocean tides and the powerful force of waves offer a consistent and reliable source of energy, making them essential components of our sustainable energy future. One of the most promising technologies for harvesting tidal energy is the use of tidal stream generators. These underwater turbines convert the kinetic energy of moving water into electrical power, with minimal environmental impact. Similarly, wave energy converters utilize the up-and-down motion of waves to generate electricity, offering a flexible and efficient way to capture wave power. Innovative designs such as oscillating water column devices and point absorbers have shown great potential in harnessing the energy from ocean waves. As technological advancements continue to improve the efficiency and cost-effectiveness of these systems, tidal and wave power are poised to play a significant role in diversifying our renewable energy portfolio. Companies and research institutions around the world are investing in the development of wave and tidal energy projects, recognizing their potential contribution to reducing carbon emissions and mitigating climate change. However, challenges such as the harsh marine environment and the high initial investment costs must be addressed to fully realize the potential of tidal and wave energy. Additionally, integrating tidal and wave power into existing energy grids requires innovative strategies to ensure a seamless and stable supply of electricity to consumers. As we navigate towards a more sustainable energy landscape, the utilization of tidal and wave power will require collaboration between governments, industry stakeholders, and the scientific community to overcome technical, regulatory, and financial barriers. With continued research and investment, tidal and wave energy can emerge as pivotal players in the global transition towards clean, renewable energy sources.

Integration of Multiple Energy Sources

As we delve into the realm of sustainable energy, the integration of multiple energy sources emerges as a pivotal aspect of modern energy production. The simultaneous utilization of various renewable energy resources offers an effective means to meet the growing demand for power while reducing reliance on fossil fuels and decreasing environmental impact. The intricate process of integrating these diverse energy sources necessitates sophisticated engineering and technological solutions that harmonize the intermittency and fluctuations inherent in renewable energy production.

The seamless integration of wind, solar, geothermal, and hydroelectric power requires a comprehensive understanding of each source's unique characteristics and operational challenges. This includes the development of advanced grid management and storage systems capable of effectively distributing and storing energy from disparate sources. Furthermore, smart grid technologies play a crucial role in balancing supply and demand in real time, enabling efficient allocation of energy from multiple sources to meet the ever-changing needs of consumers and industries.

A key consideration in the integration of multiple energy sources is the geographic and climatic diversity of energy production sites. Wind farms thrive in windy regions, while solar panels reap the most benefit in areas with abundant sunlight. Geothermal plants are situated near geological hotspots, and hydroelectric dams harness the power of flowing rivers. Coordinating the output of these varied energy sources necessitates intelligent planning and the establishment of interconnected energy infrastructures to facilitate the transfer and distribution of electricity across different regions.

Moreover, combining renewable energy sources offers resilience against external factors impacting energy generation. In instances where one source experiences reduced productivity due to adverse weather conditions or seasonal variations, the collective output from other sources can compensate for the shortfall, ensuring a consistent and reliable energy supply. By leveraging the complementary nature of these energy sources, we can mitigate the challenges associated with intermittent energy generation and achieve a more stable and secure energy grid.

The integration of multiple energy sources also contributes to significant economic and environmental benefits. Diversifying the energy mix reduces the overall vulnerability of the energy system to price fluctuations and supply disruptions, fostering greater energy security and stability. Furthermore, the increased deployment of renewable energy technologies stimulates job creation and investment in local communities, fueling economic growth and bolstering energy independence.

In conclusion, the integration of multiple energy sources represents a fundamental pillar in the transition towards sustainable and resilient energy systems. Through strategic interconnection and optimization of renewable energy resources, we can forge a cleaner, more reliable energy landscape that not only addresses our current energy needs but also safeguards the well-being of future generations.

Challenges and Solutions in Energy Harnessing

The integration of multiple energy sources presents diverse challenges that require innovative solutions for effective harnessing. A major challenge lies in the variability and intermittency of renewable energy sources such as solar and wind power. The unpredictable nature of these resources demands advanced energy storage and grid management systems to ensure a consistent power supply. Additionally, the geographical distribution of renewable energy sources necessitates long-distance transmission infrastructure, requiring substantial investment and planning.

Furthermore, the environmental impact of energy harnessing must be carefully managed. The development of large-scale renewable energy projects can lead to land use conflicts, habitat disruption, and visual impacts. Balancing energy production with environmental conservation remains a critical challenge, prompting the need for sustainable siting and development practices.

Addressing the technical aspects, the optimization of energy conversion and storage technologies is essential for maximizing efficiency and reducing costs. Research and development efforts are crucial in enhancing the performance of renewable energy systems, integrating smart grid technologies, and advancing energy storage solutions such as battery technologies and thermal storage methods.

Moreover, the socio-economic implications of energy harnessing cannot be overlooked. The transition to a sustainable energy landscape requires considerations for workforce training, job creation, and economic development in regions impacted by the shift from traditional energy sources to renewables. Social acceptance and community engagement play pivotal roles in overcoming resistance to new energy projects and fostering a supportive environment for sustainable energy initiatives.

Innovative financing mechanisms and policy frameworks are imperative for overcoming barriers to renewable energy deployment. Creating conducive regulatory environments, implementing incentive programs, and establishing supportive policies can accelerate the adoption of clean energy technologies, stimulate private investments, and drive technological innovation.

As we navigate the complex landscape of energy harnessing, collaborative efforts among government agencies, industry stakeholders, research institutions, and communities are vital. Knowledge sharing, interdisciplinary collaboration, and international cooperation can facilitate the exchange of best practices, drive standardization, and promote the global advancement of sustainable energy solutions.

Overall, addressing the multifaceted challenges in energy harnessing demands a holistic approach encompassing technological innovation, environmental stewardship, social responsibility, and favorable policy frameworks. By surmounting these challenges with integrated solutions, the path towards a resilient, low-carbon energy future becomes both feasible and imperative for the sustainable development of our planet.

Future Directions in Global Energy Policies

The future of global energy policies must address the pressing need for sustainability, efficiency, and accessibility. As we move towards a more interconnected and renewable energy-focused world, it becomes crucial to outline effective strategies for shaping energy policies on a global scale. The shift towards renewable sources such as solar, wind, and hydroelectric power requires careful consideration and planning.

One critical aspect involves international collaboration in setting standards and regulations that promote eco-friendly practices, reduce emissions, and foster technological innovation. This cooperation can facilitate the sharing of best practices, enabling countries to learn from one another's successes and challenges.

Moreover, future energy policies should underscore the importance of investing in research and development to enhance the efficiency of renewable technologies and explore emerging energy sources. This could involve incentivizing private investment, fostering public-private partnerships, and offering grants for innovative projects. By nurturing technological advancements, governments can further support the transition towards sustainable energy generation and consumption.

In addition, there is a growing emphasis on decentralization and democratization of energy production. Communities and individuals are increasingly becoming active participants in creating their own renewable energy, thereby reshaping the traditional energy landscape. Policies that support local energy initiatives, grid modernization, and energy storage systems play an essential role in this transition. They empower communities to become self-sufficient and resilient in their energy needs.

Furthermore, the integration of smart grid systems and digital technologies offers tremendous potential for optimizing energy distribution and consumption. Future energy policies should prioritize the development of smart infrastructure that enables real-time monitoring, demand response, and efficient energy use. By embracing digitalization, governments can create more flexible and responsive energy systems that adapt to dynamic energy demands.

Lastly, a holistic approach to global energy policies must also encompass socio-economic factors, ensuring that the transition to sustainable energy is inclusive and beneficial for all. This involves addressing issues of energy poverty, providing access to clean energy in underserved regions, and fostering job creation in the renewable energy sector. By prioritizing social equity and inclusivity, governments can build a more sustainable and ethical energy future.

In conclusion, shaping future global energy policies necessitates a multifaceted approach that incorporates technological innovation, international collaboration, community empowerment, and social responsibility. By setting clear and ambitious targets, investing in cutting-edge solutions, and fostering a culture of sustainability, we can pave the way towards a world powered by clean, equitable, and resilient energy.

Chapter 13

The Role of Technology in Decoding Nature

Technological Innovations in Natural Studies

Throughout history, technological innovations have played a pivotal role in advancing our understanding of ecological and biological systems. The evolution of technology has revolutionized the way we observe, analyze, and interpret nature, allowing us to delve deeper into the intricate complexities of the natural world. From early tools such as binoculars and compasses to cutting-edge satellite imagery and bioacoustic monitoring devices, each technological leap has propelled us towards a greater comprehension of the interconnected web of life. The integration of advanced sensors, data analytics, and artificial intelligence has ushered in a new era of precision and accuracy in ecological research, enabling scientists to uncover patterns and interactions that were previously concealed from our view. Remote sensing technologies and satellite imagery have expanded our capacity to monitor large-scale environmental changes, providing a comprehensive perspective of ecosystems across vast regions. These innovations have enhanced our ability to track wildlife populations, detect changes in land cover, and assess the impacts of human activities on the natural landscape. Moreover, the advent of bioacoustic monitoring has given us the capability to listen to nature's whispers, deciphering the acoustic signatures of diverse species and gaining insights into their behaviors, migrations, and communication patterns. As we harness the power of machine learning and artificial intelligence in ecological studies, we are poised to unlock intricacies that were once beyond the grasp of traditional observation methods. By leveraging these advanced technologies, researchers can process massive volumes of ecological data, identify hidden correlations, and forecast environmental trends with unprecedented precision. The integration of Geographic Information System (GIS) technology has empowered us to map and visualize the language of landscapes, facilitating spatial analysis and modeling of ecological processes. Drones have emerged as invaluable tools for capturing high-resolution imagery and conducting aerial surveys, offering new perspectives on habitats, topographies, and ecological dynamics. The synergy of Internet of Things (IoT) devices in ecological monitoring has elevated our real-time understanding of environmental parameters, fostering a more proactive approach to conservation and management. However, alongside the remarkable progress, it is essential to acknowledge the challenges and limitations associated with the application of technology in natural studies. From data privacy concerns to the ethical use of advanced surveillance, navigating the ethical landscape of technological innovations demands thoughtful consideration and responsible stewardship. Nonetheless, as technology continues to evolve at a rapid pace, the future holds immense promise for novel advancements that will revolutionize our capacity to decode the intricate tapestry of nature.

Remote Sensing and Satellite Imagery: Expanding Our View

The advent of remote sensing and satellite imagery has revolutionized the realm of environmental studies, offering a bird's-eye view of our planet's diverse landscapes and ecosystems. These advanced technologies have empowered scientists to transcend geographical boundaries and access remote or inaccessible regions, thus enhancing our understanding of natural processes and patterns on a global scale. Through the deployment of satellites equipped with state-of-the-art sensors and cameras, researchers can capture high-resolution images and data, enabling them to monitor changes in land cover, vegetation health, and atmospheric conditions with unprecedented accuracy. Remote sensing plays a pivotal role in mapping out localized phenomena such as deforestation, urban expansion, agricultural trends, and natural disasters, providing valuable insights for conservation efforts and sustainable management practices. By harnessing the power of multispectral and hyperspectral imaging, scientists can analyze the reflective properties of different surfaces and materials, discerning subtle variations in soil composition, water quality, and biodiversity. Moreover, the integration of remote sensing data with geographic information systems (GIS) has facilitated the development of detailed maps, digital elevation models, and spatial databases, which serve as indispensable tools for environmental planning, resource assessment, and ecological modeling. The application of remote sensing techniques extends beyond Earth, as it also enables us to explore celestial bodies and phenomena in the cosmos. From monitoring climate change impacts to tracking wildlife populations and migratory patterns, remote sensing and satellite imagery continue to unlock new frontiers in ecological research, reaffirming their status as indispensable assets in the pursuit of deciphering nature's intricate language.

Bioacoustic Monitoring: Listening to Nature's Whispers

Bioacoustic monitoring, also known as acoustic ecology, entails the use of sound recordings to analyze and interpret the acoustic behaviors of various organisms within their natural environments. This cutting-edge approach offers unique insights into the intricate communication systems present in ecosystems, shedding light on the hidden language of nature. By capturing the subtle nuances of vocalizations, calls, and other acoustic signals, researchers can unravel a wealth of information about ecological patterns, species interactions, and environmental changes.

The utilization of bioacoustics in ecological studies has revolutionized our understanding of wildlife behavior and ecosystem dynamics. Through advanced recording equipment and sensitive microphones, scientists can capture and document a diverse array of sounds, ranging from the melodic songs of birds to the rhythmic choruses of amphibians and insects. These recordings serve as valuable data sources, providing critical information for conservation efforts, habitat management, and biodiversity assessments.

Moreover, bioacoustic monitoring plays a pivotal role in detecting subtle shifts in natural soundscapes due to anthropogenic activities, climate change, and habitat degradation. By establishing baseline acoustic profiles, researchers can monitor changes over time, identify potential threats to biodiversity, and assess the effectiveness of conservation measures. The interdisciplinary nature of bioacoustics integrates elements of biology, ecology, acoustics, and technology, fostering novel collaborations and innovative research methodologies.

Furthermore, advancements in machine learning and automated signal processing have facilitated the analysis of extensive acoustic datasets, enabling researchers to discern meaningful patterns and decipher complex environmental codes. By employing sophisticated algorithms, bioacoustic experts can classify species-specific vocalizations, quantify population dynamics, and even identify the presence of elusive or endangered species based on their distinct acoustic signatures.

As technology continues to evolve, the potential applications of bioacoustic monitoring are expanding rapidly. From eco-sensitive urban planning and environmental soundscape design to non-invasive monitoring of sensitive ecosystems, the implications of bioacoustics are far-reaching. This burgeoning field holds immense promise for deepening our connection with the natural world, fostering conservation awareness, and preserving the rich tapestry of sounds that define our planet's unparalleled biodiversity.

Data Analytics and Pattern Recognition: Deciphering Environmental Codes

In the realm of environmental science, the application of data analytics and pattern recognition has revolutionized our capacity to decode the intricacies of natural phenomena. Through advanced computational techniques, scientists can now sift through vast amounts of environmental data to identify patterns, trends, and correlations that were previously indiscernible. This innovative approach enables researchers to uncover hidden codes within ecological systems, leading to a deeper understanding of the complexities of our natural world. By harnessing the power of data analytics, we can unravel the intricate web of interactions between different components of ecosystems, from climate patterns to species behaviors. Moreover, the identification of these patterns allows us to predict and mitigate environmental changes, thereby contributing to the preservation of biodiversity and the sustainability of our planet. One of the most compelling aspects of utilizing data analytics and pattern recognition in environmental studies is the ability to integrate diverse datasets from various sources. By amalgamating information from satellite imagery, ground-based sensors, biological surveys, and historical weather records, researchers can construct comprehensive models that encapsulate the multifaceted nature of environmental processes. Furthermore, the emergence of machine learning algorithms has augmented the efficacy of pattern recognition, enabling the identification of subtle correlations and anomalies that elude traditional analytical methods. This newfound capability holds profound implications for ecological conservation and management, as it empowers scientists to make informed decisions based on comprehensive and nuanced insights. As technology continues to advance, the field of environmental science stands at the precipice of an era where the deciphering of environmental codes will be integral to addressing global challenges such as climate change, habitat loss, and biodiversity decline. Embracing this paradigm shift in natural studies will not only deepen our comprehension of the intricate language spoken by our planet but also pave the way for sustainable coexistence with nature, fostering a harmonious relationship between humanity and the environment.

Machine Learning Applications in Flora and Fauna Research

Machine learning has rapidly emerged as a transformative tool in the realm of environmental studies, particularly in enhancing our understanding of the behavior and characteristics of flora and fauna. By leveraging algorithms that can learn from and make predictions based on data, researchers have been able to uncover profound insights into the intricate interactions within ecosystems. At the forefront of this technological revolution is the application of machine learning in species identification and classification. Gone are the days of painstakingly cataloging and identifying species by hand; with machine learning models trained on vast datasets of plant and animal features, accurate and efficient species recognition has become a reality. This not only expedites the process of biodiversity surveys but also enables tracking changes in species distribution over time, vital for conservation efforts. Furthermore, machine learning models have proven invaluable in predicting ecological phenomena such as the impact of climate change on migration patterns or the spread of invasive species. Through pattern recognition and analysis of large-scale environmental data, these models can forecast potential shifts in ecosystems, providing early warnings for conservationists and policymakers. In addition to species identification and ecological forecasting, machine learning has empowered researchers to delve deeper into understanding complex ecological processes. By processing enormous volumes of environmental sensor data, machine learning algorithms can unveil previously unnoticed correlations and patterns, shedding light on intricate ecosystem dynamics. Whether it is delineating habitat preferences of endangered species or modeling food web interactions, machine learning offers a powerful lens through which we can interpret and protect the natural world. Moreover, the integration of machine learning with remote sensing data has opened up new vistas in landscape ecology. From mapping vegetation cover to monitoring wildlife populations, machine learning algorithms can derive invaluable insights from satellite and aerial imagery. This not only circumvents the need for labor-intensive field surveys but also enables a comprehensive and continuous assessment of environmental changes at various spatial and temporal scales. However, it is essential to acknowledge the ethical and methodological considerations in the utilization of machine learning in ecological research. Ensuring unbiased and representative training data, addressing algorithmic biases, and maintaining transparency in model development are pivotal in upholding the integrity of research outcomes. As the field continues to evolve, interdisciplinary collaborations between ecologists, data scientists, and ethicists will be crucial in harnessing the full potential of machine learning for the betterment of our planet.

GIS Technology: Mapping the Language of Landscapes

Geographic Information System (GIS) technology has revolutionized the way we perceive and understand the intricate language of landscapes. By integrating spatial data with advanced analytical tools, GIS allows us to unravel the hidden narratives encoded in the Earth's topography. Through the utilization of satellite imagery, aerial photographs, and ground-based sensors, GIS provides a comprehensive framework for mapping and interpreting the diverse elements of landscapes. This powerful tool enables researchers to analyze land cover, terrain characteristics, vegetation patterns, and hydrological features with unprecedented precision. Furthermore, GIS facilitates the assessment of environmental changes over time, offering valuable insights into the dynamic interactions between natural systems and human activities. By overlaying multiple layers of geographic information, GIS technology empowers us to elucidate the complex relationships between various components of ecosystems, enabling a holistic comprehension of landscape dynamics. The integration of spatial analysis and modeling within GIS platforms enhances our capacity to forecast ecological trends, assess biodiversity hotspots, and mitigate environmental risks. Moreover, GIS technology plays a pivotal role in conservation efforts, aiding policymakers and environmental organizations in the effective management of natural resources and protected areas. By mapping habitat distributions, migration corridors, and ecological connectivity, GIS contributes to the design of sustainable land-use strategies and biodiversity conservation plans. With its ability to visualize, interpret, and communicate geographic data, GIS serves as a fundamental tool for decision-making processes related to land management, urban planning, and disaster preparedness. As we continue to delve deeper into the language of landscapes, GIS technology remains an indispensable asset in our quest to comprehend and preserve the intricate tapestry of the natural world.

Drone Technology: New Perspectives on Old Mysteries

The emergence of drone technology has revolutionized our ability to gain new insights and perspectives on age-old mysteries within the natural world. Drones, equipped with advanced cameras and sensors, have transcended conventional limitations and provided researchers with a bird's eye view of landscapes and ecosystems that were previously inaccessible. This technological innovation has opened up unparalleled opportunities for exploring and understanding the intricate language of nature. By capturing high-resolution imagery and videos, drones enable scientists and conservationists to document environmental changes, migratory patterns, and habitat variations in a non-intrusive manner.

Moreover, drone technology has proven to be instrumental in wildlife monitoring and management. With their ability to access remote and rugged terrains, drones facilitate the tracking of elusive species and provide real-time data on behavior and population dynamics. This enhanced capability has transformed the way researchers approach ecological studies, enabling them to gather comprehensive information without disrupting the delicate balance of natural ecosystems. Furthermore, in the realm of archaeology and historical ecology, drones have unveiled hidden archaeological sites, ancient ruins, and geoglyphs, shedding light on civilizations long lost to time. The aerial perspective offered by drones has allowed experts to identify and analyze previously unknown features, revolutionizing our understanding of ancient cultures and the evolution of landscapes.

Complementing these advancements, drone technology has facilitated rapid response to environmental disasters and ecological threats. Equipped with thermal imaging and infrared capabilities, drones aid in identifying fire outbreaks, monitoring deforestation activities, and assessing the impact of natural calamities on ecosystems. This proactive approach not only helps in mitigating the repercussions of such events but also contributes to the formulation of robust conservation and restoration strategies. Additionally, in marine biology and oceanography, drones have enabled the monitoring of marine life, coral reefs, and coastal ecosystems with unprecedented precision, unveiling hitherto unseen aspects of underwater habitats and species behavior.

As with any technological advancement, ethical considerations and privacy concerns surrounding the use of drones in nature must be carefully addressed. Regulations and guidelines need to be established to ensure that the deployment of drones aligns with ethical research practices and safeguards the welfare of wildlife and ecosystems. It is imperative to strike a balance between leveraging the potential of drone technology for scientific exploration and ensuring its responsible and sustainable utilization. As we continue to harness the power of drone technology in decoding nature's mysteries, it is essential to remain mindful of the ethical dimensions associated with this transformative tool.

The Integration of IoT in Ecological Monitoring

As the field of ecological monitoring continues to evolve, the integration of Internet of Things (IoT) technologies has emerged as a transformative force. IoT refers to the interconnected network of devices embedded with sensors, software, and other technologies for the purpose of collecting and exchanging data. When applied to ecological monitoring, IoT enables researchers and environmentalists to gather real-time, granular data across various ecosystems and habitats, revolutionizing our understanding of natural processes.

One key aspect of IoT in ecological monitoring is its ability to provide continuous, unobtrusive observation of wildlife behavior and environmental dynamics. Sensor-equipped devices, such as camera traps and acoustic sensors, can capture valuable information on species activity patterns, population dynamics, and habitat usage without disturbing the natural ecosystem. This non-invasive approach allows for a more comprehensive and accurate assessment of ecological systems.

Furthermore, the integration of IoT facilitates the remote collection and transmission of environmental data, eliminating the need for frequent physical intervention in the field. Automated weather stations, soil moisture sensors, and water quality monitors can continuously relay vital environmental parameters to centralized databases, enabling real-time analysis and interpretation. This capability not only enhances the efficiency of ecological monitoring efforts but also reduces human impact on fragile ecosystems.

In addition to on-site data collection, IoT applications extend to the realm of habitat restoration and conservation. By deploying IoT-enabled devices in degraded areas, conservationists can monitor the effectiveness of restoration initiatives, track changes in vegetation cover, and assess the return of native fauna. Moreover, IoT plays a crucial role in early detection and rapid response to environmental threats, such as pollution or illegal activities, aiding in the preservation of precious natural resources.

Despite its numerous benefits, the integration of IoT in ecological monitoring also presents challenges and considerations. Data security and privacy issues, network connectivity in remote locations, and the potential for sensor malfunctions are among the primary concerns that require careful mitigation strategies. Additionally, the ethical use of IoT technology in wildlife research and conservation must be thoughtfully addressed to ensure minimal disruption to natural ecosystems and species.

Looking ahead, the future of IoT in ecological monitoring holds immense promise. Advancements in sensor technology, wireless communication, and data analytics will further enhance the precision and scope of ecological observations. By harnessing the power of IoT, we stand poised to unlock deeper insights into the complex interactions of natural systems, ultimately contributing to informed decision-making and sustainable environmental stewardship.

Challenges and Limitations of Technology in Natural Observations

As we delve deeper into the era of technological advancements in decoding nature, it is essential to acknowledge the challenges and limitations that accompany this progress. One of the primary challenges lies in the potential bias introduced by technological tools in ecological monitoring. The reliance on specific sensors or remote devices may inadvertently skew the data collected, leading to a distorted understanding of natural phenomena. Additionally, the sheer volume of data generated by modern technological systems presents a significant challenge in itself. The collection and processing of an overwhelming amount of information often surpasses human capabilities, demanding advanced computing resources and expertise. Another critical limitation involves the accessibility and affordability of high-tech equipment in various geographical regions. Many areas with rich biodiversity and ecological significance lack the infrastructure and resources necessary for the widespread implementation of cutting-edge technologies, resulting in unequal representation and gaps in our understanding of global ecosystems. Furthermore, technological interventions can disrupt natural habitats and behaviors, raising ethical concerns about the potential interference with the ecological balance. Additionally, the rapid pace of technological evolution poses another challenge, as conservation efforts struggle to keep up with the ever-changing landscape of tools and methodologies. Despite these challenges, it is crucial to recognize that technology is a tool crafted by humans, and its limitations are not insurmountable. By acknowledging these challenges and actively seeking solutions, such as standardizing data collection protocols and promoting inclusivity in technological access, we can navigate these obstacles and harness the full potential of technology in our quest to decode the language of nature.

Future Trends in Technology for Environmental Communication

As we stand at the precipice of a new era in environmental communication, it is imperative to explore the potential trajectory of technological advancements in this domain. The convergence of various cutting-edge technologies promises to revolutionize our understanding and interpretation of nature's intricate language. One notable trend on the horizon is the proliferation of quantum computing applications in environmental research. The unparalleled processing power of quantum computers has the potential to unravel complex environmental patterns and phenomena that are currently beyond the scope of conventional computing capabilities. Furthermore, the integration of artificial intelligence (AI) and machine learning algorithms into environmental monitoring systems is poised to elevate our capacity to interpret and analyze vast amounts of data with unprecedented accuracy and efficiency. By harnessing AI for predictive modeling and adaptive decision-making, we can anticipate ecological changes and mitigate potential risks with greater precision. Alongside the advancement of AI, the emergence of blockchain technology offers promise in securing and transparently sharing ecological data across decentralized networks. This distributed ledger system not only enhances data integrity but also fosters collaborative efforts in environmental conservation and research on a global scale. Another trend gaining momentum pertains to the refinement of biometric sensors and IoT devices for real-time environmental monitoring. These compact yet powerful instruments enable continuous data collection, allowing researchers to track subtle fluctuations in natural ecosystems and wildlife behaviors. Additionally, advancements in nanotechnology are envisioned to introduce nano-scale sensors capable of probing environmental parameters at an unprecedented level of granularity. As we look ahead, the fusion of advanced sensor networks with big data analytics holds immense potential in uncovering hidden ecological patterns and interconnections. The advent of genomic editing tools such as CRISPR-Cas9 also presents an intriguing frontier in environmental communication. By manipulating the genetic traits of key species, scientists envision the restoration of ecosystems and the mitigation of environmental degradation. Furthermore, the development of low-orbit satellite constellations promises enhanced global coverage for environmental monitoring, enabling comprehensive observation of remote and inaccessible regions. Lastly, the anticipated growth of quantum communication technologies signifies a shift towards ultra-secure and high-fidelity transmission of environmental data across vast distances. By embracing these future trends, we are poised to embark on a transformative journey in decoding the language of nature and cultivating a deeper harmony with our planet.

Chapter 14

Unraveling the Language of the Cosmos

Cosmic Communication

Cosmic communication has long held a profound fascination for humanity, encompassing various attempts to decipher and comprehend the language of the cosmos. The concept of cosmic communication revolves around the transmission and reception of signals, messages, and interactions that originate from celestial bodies, phenomena, or interstellar spaces. This fundamental facet of astrophysical inquiry seeks to unravel the complexities of cosmic languages, spanning astronomical events, extraterrestrial intelligence, and the interconnected web of celestial entities. At its core, cosmic communication encompasses both the conceptualization and theoretical investigation of diverse forms of potential transmissions from the expansive realms beyond our planet. It is imperative to recognize that the conjecture surrounding cosmic communication extends far beyond mere linguistic expressions, delving into the essence of information exchange, encoding mechanisms, and the potential dialogues that may exist between cosmic entities. The central pursuit of understanding and interpreting such communications lies at the crossroads of astronomy, astrobiology, and the broader interdisciplinary tapestry of space exploration. In essence, the study of cosmic communication employs a multidimensional framework that fuses elements of physics, information theory, and advanced signal processing methodologies. It seeks to discern patterns, regularities, and anomalies within the fabric of cosmic emissions, with the ultimate objective of unveiling the underlying message they might convey. This chapter endeavors to navigate the theoretical underpinnings that underlie cosmic communication, shedding light on historical perspectives, contemporary paradigms, and emerging frontiers in our quest to decode the cryptic whispers of the universe.

Historical Perspectives on Celestial Languages

Throughout the ages, human civilization has looked to the skies for inspiration and guidance. From ancient cultures to the present day, the celestial bodies have captivated our imagination and curiosity, leading to numerous theories and beliefs about their significance and potential messages they might hold. Historically, diverse civilizations have interpreted the celestial lights as omens, divine pronouncements, or even direct forms of communication from the gods. The alignments of stars and movements of planets were often seen as secret codes carrying profound meaning. In ancient Mesopotamia, the Babylonians meticulously recorded celestial events, developing an early form of astrology as a means of divination and understanding the will of the deities. Similarly, ancient Egyptian civilization held celestial bodies in high regard, associating them with their pantheon of gods and using their positions to govern various religious rituals and ceremonies. The Greeks also made significant advances in understanding the cosmos, with philosophers like Aristotle and Ptolemy laying foundations for the geocentric model of the universe, based on the belief that celestial bodies conveyed divine principles and cosmic order. Moving through the medieval era, the study of celestial phenomena transitioned into the realms of astronomy and astrology, where scholars and practitioners sought to comprehend the language of the heavens in shaping terrestrial events and human destinies. The Renaissance and Age of Enlightenment marked a pivotal shift, as the focus moved towards empirical observation and scientific inquiry, leading to remarkable discoveries by pioneers such as Copernicus, Galileo, and Kepler, which revolutionized humanity's understanding of the cosmos. As scientific knowledge expanded, so did our perspectives on celestial languages, gradually transcending notions of mysticism and superstition. By tracing these historical threads, we gain crucial insights into the profound impact of celestial bodies on human culture and intellectual evolution, setting the stage for contemporary explorations into unraveling the enigmatic messages encoded within the cosmic expanse.

Astrophysical Signals and Their Interpretations

Astrophysical signals are the enigmatic whispers of the cosmos, echoing through the vast expanse of space. These signals manifest in various forms, from electromagnetic waves to gravitational waves, each carrying crucial information about the nature of celestial bodies and their interactions. The study of astrophysical signals represents a fundamental component of modern cosmology, offering profound insights into the origins and evolution of our universe.

Interpreting these signals demands a multifaceted approach, integrating principles from physics, mathematics, and astronomy. One of the most pervasive sources of astrophysical signals is the electromagnetic spectrum, encompassing visible light, radio waves, microwaves, infrared radiation, ultraviolet rays, X-rays, and gamma rays. Each band of the electromagnetic spectrum unveils unique aspects of cosmic phenomena, enabling astronomers to peer into the heart of distant galaxies, observe the birth of stars, and unravel the mysteries of black holes.

Moreover, gravitational waves, first theorized by Albert Einstein, have emerged as a transformative tool for probing the fabric of spacetime itself. These ripples in the continuum carry information about cataclysmic events such as collisions between massive celestial bodies, providing scientists with a new lens to explore the universe's most violent and dynamic processes.

The interpretation of astrophysical signals often involves intricate data analysis and computational modeling. By leveraging mathematical techniques, researchers can extract valuable information from raw signal data, discerning patterns that elucidate the underlying physical mechanisms at play. Furthermore, advanced algorithms and machine learning methodologies have empowered astronomers to sift through vast datasets, identifying subtle nuances in astrophysical signals that might otherwise escape detection.

In the quest to interpret astrophysical signals, interdisciplinary collaboration plays a pivotal role. Astrophysicists, mathematicians, and data scientists work hand in hand, applying diverse analytical tools to unravel the complex tapestry of signals emanating from cosmic phenomena. The synergy between different disciplines fosters a holistic understanding of astrophysical processes and enables the development of innovative methods for decoding the intricacies of celestial communication.

The ongoing pursuit to decode astrophysical signals holds profound implications for our comprehension of the universe. By honing our ability to interpret these signals, we embark on a journey to unveil the cosmic narrative, shedding light on the ancient conversations that resonate throughout the cosmos.

The Role of Mathematics in Deciphering Cosmic Patterns

Mathematics plays an indispensable role in unraveling the enigmatic language of the cosmos. Through advanced mathematical models and theories, scientists have been able to decode and comprehend the intricate patterns and phenomena observed in the celestial realm. The fundamental principles of mathematics, such as calculus, geometry, and algebra, form the basis for understanding the behavior of celestial bodies and the complex interplay of forces that govern the universe. Mathematical equations provide a language through which cosmic patterns can be described, quantified, and ultimately deciphered.

The utilization of mathematical tools such as differential equations and computational algorithms has enabled researchers to analyze vast sets of astronomical data, identifying regularities and anomalies that offer crucial insights into the underlying structure of the cosmos. From the trajectories of celestial objects to the gravitational interactions between galaxies, mathematics serves as the essential framework for interpreting these celestial phenomena.

Moreover, mathematical concepts such as Fourier analysis and statistical methods are instrumental in discerning recurring patterns or signals emanating from deep space, contributing to the identification of cosmic events and astronomical objects. The application of probability theory and stochastic processes further facilitates the assessment of uncertainties and fluctuations inherent in cosmic observations, shedding light on the probabilistic nature of cosmic phenomena.

The integration of mathematics with astrophysics has led to groundbreaking discoveries, including the prediction and subsequent detection of gravitational waves, the formulation of cosmological models describing the evolution of the universe, and the identification of exoplanetary systems through transit analysis. These achievements highlight the indispensable role of mathematics as a universal language in decoding cosmic patterns and gaining profound insights into the nature of the cosmos.

As technology continues to advance, the synergy between mathematics and astronomy has grown even stronger. High-performance computing and simulation techniques allow for the numerical modeling of complex astrophysical processes, enabling scientists to replicate and study cosmic events under diverse conditions. Furthermore, the development of data-driven approaches, machine learning algorithms, and artificial intelligence has revolutionized the analysis of astronomical datasets, leading to the discovery of new celestial phenomena and the characterization of previously unexplored aspects of the universe.

In essence, the marriage of mathematics and astronomy stands as a testament to human ingenuity and intellectual curiosity, unlocking the secrets of the cosmos through a harmonious blend of abstract mathematical reasoning and empirical astronomical observations. The profound impact of mathematics in deciphering cosmic patterns transcends the boundaries of earthly knowledge, guiding humanity towards a deeper understanding of the vast and wondrous universe that surrounds us.

Technological Advancements and Space Explorations

Space exploration has always been at the forefront of human curiosity and technological advancement. Over the years, significant progress has been made in developing advanced technologies that allow mankind to venture beyond Earth's atmosphere and explore the cosmos. Technological advancements in space exploration have not only expanded our understanding of the universe but also revolutionized various scientific disciplines on Earth. The development of powerful telescopes, probes, and space vehicles has enabled scientists to study celestial bodies, cosmic phenomena, and environmental conditions in outer space. These tools have provided unprecedented insights into the nature of the universe, enhancing our knowledge of distant planets, stars, galaxies, and black holes. Furthermore, space explorations have facilitated groundbreaking discoveries related to fundamental concepts such as gravity, relativity, and the origins of the cosmos. As technology continues to evolve, new frontiers in space exploration are constantly being pursued. Innovations in propulsion systems, spacecraft design, and communication technologies are enabling humans to target more ambitious space missions, including crewed missions to Mars, deep-space explorations, and the establishment of lunar habitats. Moreover, robotic missions to distant celestial bodies have paved the way for sustainable and cost-effective space exploration, with the potential for mining resources from asteroids and other planetary bodies. The collaborative efforts of space agencies, scientific organizations, and private space companies are driving the development of cutting-edge technologies, fostering international cooperation and promoting the peaceful exploration of outer space. Such endeavors are reshaping our understanding of the universe and inspiring new generations of scientists, engineers, and explorers to push the boundaries of what is possible. The marriage of technological advancements and space exploration continues to kindle the spirit of discovery, pushing humanity to unlock the mysteries of the cosmos and realize the dream of venturing into the unknown depths of space.

Analyzing the Influence of Cosmic Events on Earth

Cosmic events have exerted a profound influence on Earth since the dawn of time. From lunar phases shaping agricultural practices to the dramatic impacts of solar flares, our planet is intricately connected to the celestial phenomena occurring in the universe. Understanding and analyzing this influence is vital for unlocking the secrets of our planet's history and predicting its future. The gravitational tug-of-war between the Moon and Earth not only causes the ebb and flow of tides but also affects geological processes such as volcanic activity and earthquake patterns. Moreover, the positioning and movement of other celestial bodies like comets and asteroids have left indelible marks on Earth's surface through their impact craters, altering the course of evolution and shaping the landscape we see today. Additionally, solar activity, including sunspots and solar flares, can cause geomagnetic storms that interact with Earth's magnetic field, affecting power grids, communication systems, and even leading to breathtaking auroras. Furthermore, cosmic radiation constantly bombards our planet, influencing mutation rates in organisms and potentially impacting climate patterns. By closely scrutinizing these influences, scientists can gain invaluable insights into Earth's past and develop more accurate models to forecast future changes. Advancements in technology, such as sophisticated telescopes, satellite observations, and interdisciplinary research approaches, have revolutionized our ability to monitor and analyze cosmic events and their impact on Earth. Integrating data from diverse fields, including astronomy, geology, meteorology, and biology, enables us to construct a comprehensive picture of how celestial occurrences shape our planet. This in-depth understanding not only enriches our knowledge of Earth's history but also provides essential information for mitigating potential risks associated with future cosmic events. As we continue our exploration of the cosmos, delving deeper into the dynamic relationships between celestial events and Earth's evolution, we are poised to unravel even more of the intricate connections that bind our planet to the broader universe.

Decoding the Message Embedded in Cosmic Microwaves

Cosmic microwaves, a form of electromagnetic radiation, hold a treasure trove of information about the early universe. These faint signals, permeating the cosmos, offer insights into the primordial conditions that prevailed shortly after the Big Bang. Within these cosmic microwaves lies an ancient language written in the fabric of space-time, waiting to be decoded by astrophysicists and cosmologists alike.

One of the fundamental pursuits in modern cosmology is to decipher the cosmic microwave background (CMB) radiation. These electromagnetic waves, originating from nearly 13.8 billion years ago, provide a snapshot of the universe when it was a mere 380,000 years old. By analyzing the minute temperature fluctuations within the CMB, scientists gain critical knowledge about the composition, evolution, and geometry of the universe on the largest scales.

Decoding the message embedded in cosmic microwaves requires intricate data analysis and sophisticated theoretical models. Through advanced observational techniques and cutting-edge technologies such as the Planck satellite and ground-based telescopes, researchers meticulously map and scrutinize the cosmic microwave background. By discerning subtle patterns and anomalies within the CMB radiation, they unearth clues about the cosmic web, dark matter, dark energy, and the seeds of structure formation.

Furthermore, the cosmic microwave background aids in verifying key components of the Big Bang theory. It offers empirical support for the inflationary model, shedding light on the rapid expansion of the universe moments after its birth. Moreover, the CMB holds the promise of unraveling the enigma of cosmic polarization, potentially corroborating the existence of gravitational waves generated during inflation.

With technological advancements and interdisciplinary collaborations, the study of cosmic microwaves continues to revolutionize our understanding of the universe's origin and evolution. The precise measurements and intricate analyses of the CMB not only deepen our knowledge of cosmology but also ignite fresh inquiries that propel scientific exploration to new frontiers. Unraveling the intricacies of the cosmic microwave background stands as a testament to humanity's insatiable curiosity and relentless pursuit of cosmic comprehension.

Interpreting the Language of Neutrinos

Neutrinos, often referred to as 'ghost particles', are elusive and incredibly tiny subatomic particles that are constantly traversing the universe. As we delve into the intricate fabric of cosmic communication, understanding the role of neutrinos becomes paramount. These enigmatic particles are generated by various celestial phenomena, such as nuclear reactions within stars, supernovae explosions, and cosmic ray interactions. What sets neutrinos apart from other particles is their weak interactions with matter, allowing them to travel vast distances through space, encountering minimal obstructions. In the realm of cosmic linguistics, neutrinos serve as carriers of invaluable information about distant astrophysical events.

The study of neutrinos encompasses different detection methods such as neutrino telescopes, where scientists can observe high-energy neutrinos produced in cosmic sources. By examining the energy spectrum and arrival direction of these neutrinos, researchers can infer details about their origins, shedding light on the nature of the processes occurring in distant celestial objects. Moreover, the detection of neutrinos can unveil vital insights into cosmic mysteries, including the formation of black holes, neutron stars, and other astronomical enigmas. The inherent properties of neutrinos enable them to act as messengers from realms previously inaccessible to human observation.

Furthermore, the interpretation of neutrino oscillations – the phenomenon where neutrinos change flavor as they propagate through space – provides a unique avenue for deciphering cosmic codes. By analyzing the oscillation patterns, scientists can discern the properties of the neutrinos' source and gain deeper comprehension of cosmic phenomena. Additionally, the investigation of neutrino astronomy contributes to interdisciplinary collaboration across astrophysics, particle physics, and cosmology, fostering a holistic approach to understanding the universe's language.

As technology continues to advance, innovative experiments such as IceCube Neutrino Observatory and The Super-Kamiokande are pushing the boundaries of our knowledge about neutrinos, offering unprecedented opportunities for unraveling the cosmic narrative. These endeavors not only expand our understanding of the universe but also present prospects for breakthrough discoveries that may revolutionize our perception of the cosmos. Through the relentless pursuit of unraveling the language of neutrinos, humanity moves closer to comprehending the profound interplay between celestial entities and the universal symphony harbored within the cosmos.

Future Prospects in Cosmic Linguistics

The gradual advancement of technology and the evolving understanding of cosmic phenomena have significantly broadened the scope of cosmic linguistics. As we look to the future, one of the most promising avenues is the potential discovery of extraterrestrial intelligence. With ongoing missions such as SETI (Search for Extraterrestrial Intelligence) scanning the cosmos for transmissions that could originate from advanced civilizations, the field of cosmic linguistics may soon face an unprecedented breakthrough.

Furthermore, the integration of artificial intelligence and machine learning algorithms into the analysis of cosmic signals holds immense promise. These technologies can aid in sifting through vast amounts of data to identify patterns or anomalies that could signify deliberate communication or natural yet unexplained cosmic phenomena. Researchers are increasingly employing these tools to analyze complex astrophysical signals and predict cosmic events, revealing a potential roadmap for the future of cosmic linguistics.

Moreover, the prospect of establishing interstellar communication using quantum entanglement has garnered attention from physicists and engineers alike. The instantaneous transmission of information over vast cosmic distances presents an enticing frontier for cosmic linguistics, enabling the possibility of real-time dialogue with extraterrestrial civilizations should they exist. The theoretical groundwork for such communication is being explored, fueling the imagination of scientists and science fiction enthusiasts alike.

Additionally, the study of cosmic languages extends beyond the search for intelligent life. With advancements in our understanding of dark matter and dark energy, cosmic linguistics may uncover new insights into the fundamental forces shaping the universe itself. By deciphering the hidden messages carried by cosmic elements, be they neutrinos, gravitational waves, or other enigmatic signals, researchers aim to unravel the mysteries of the cosmos and gain a deeper comprehension of the universe at its most fundamental level.

Finally, future prospects in cosmic linguistics also entail collaborative efforts across diverse scientific disciplines. From astrophysics to anthropology, integrating various fields of study will be crucial in fostering a holistic approach to understanding the language of the cosmos. This interdisciplinary synergy has the potential to yield groundbreaking discoveries, fostering a new era of cosmic exploration and enlightenment.

Conclusion: Integrating Cosmic Codes into Modern Science

The exploration of cosmic codes has opened up a realm of unprecedented scientific potential, offering insights that transcend the boundaries of our current understanding. As we delve deeper into the intricacies of celestial communication, it becomes increasingly evident that the integration of cosmic codes into modern science carries profound implications for various disciplines. By embracing and deciphering the enigmatic messages embedded in the cosmos, we stand to revolutionize our comprehension of the universe and propel scientific progress to unparalleled heights.

One of the most compelling aspects of integrating cosmic codes into modern science is the opportunity to gain a deeper awareness of our cosmic origins. By studying the intricate language of the cosmos, we can unravel the mysteries surrounding the formation of galaxies, stars, and planets, shedding light on the fundamental processes that have shaped the universe over billions of years. This illuminating journey into the depths of cosmic communication not only enhances our cosmic perspective but also establishes a framework for understanding our place within the vast expanse of the cosmos.

Moreover, the integration of cosmic codes has the potential to yield groundbreaking advancements in astrophysics and cosmology. The ability to interpret and comprehend the celestial signals emanating from distant corners of the universe provides astrophysicists with invaluable data that can drive innovative discoveries and theoretical breakthroughs. From elucidating the nature of dark matter and dark energy to unraveling the mysteries of black holes and neutron stars, the integration of cosmic codes into modern science fuels a relentless pursuit of knowledge that transcends the confines of our terrestrial domain.

Furthermore, the profound implications of integrating cosmic codes extend beyond the realms of astrophysics, permeating fields such as mathematics, technology, and even philosophy. The utilization of advanced mathematical algorithms for decoding cosmic patterns and signals not only deepens our understanding of the universe but also enhances mathematical frameworks that can be applied to diverse scientific domains. Similarly, technological innovations driven by cosmic communication research have the potential to catalyze transformative developments in space exploration, telecommunications, and signal processing, thereby exerting a far-reaching influence on humanity's technological evolution.

As modern science continues to assimilate cosmic codes into its repertoire, the interdisciplinary nature of this endeavor becomes increasingly apparent. In the quest to unlock the profound messages encoded within the fabric of the cosmos, collaborations between scientists from varied fields converge to form a tapestry of knowledge that transcends traditional disciplinary boundaries. Through these collaborative endeavors, the synthesis of diverse perspectives and expertise enriches the collective understanding of cosmic communication, fostering a dynamic ecosystem of scientific inquiry that drives monumental progress in unraveling the mysteries of the universe.

In essence, the integration of cosmic codes into modern science marks a pivotal juncture in humanity's quest for knowledge, symbolizing an enduring commitment to unraveling the enigmatic language of the cosmos. By weaving the intricate threads of celestial communication into the fabric of scientific exploration, we embark on a transformative journey that holds the promise of expanding our horizons, catalyzing innovation, and illuminating the profound mysteries that transcend the boundaries of our earthly existence.

Chapter 15

Implications for Climatology: Predicting Future by the Past

Bridging Past Climates with Future Forecasts

Historical climatic events offer a crucial lens through which we can comprehend the intricacies and potential trajectories of future climatology. By delving into ancient climate patterns, we gain valuable insights into the complex mechanisms shaping our planet's environmental processes. The link between historical climate data and contemporary trends presents an invaluable opportunity to fortify our predictive capabilities, enabling us to anticipate and potentially mitigate the impacts of impending environmental shifts. Moreover, understanding the interconnectedness of past and present climates serves as a foundational pillar for developing adaptive strategies in a rapidly evolving world. Through an exploration of historical climate patterns and their continuity into the current era, we can illuminate the interplay of natural forces and anthropogenic influences, fostering a deeper comprehension of the intricate web of factors steering our planet's climate. This section aims to elucidate the profound importance of historical climatic events as a cornerstone for formulating comprehensive and reliable forecasts, thereby empowering individuals, communities, and policymakers to proactively address the challenges posed by climate change. Embracing the knowledge gleaned from the past allows us to not only appreciate the magnitude of environmental transformations but also empowers us to forecast potential future scenarios with increased precision and foresight.

Historical Climate Patterns and Current Trends

Understanding historical climate patterns and their influence on current trends is critical for comprehending the complex dynamics of our planet's climate system. By examining past climate data, scientists can identify recurring patterns, fluctuations, and anomalies that have shaped the Earth's climatic history. This retrospective analysis provides invaluable insights into the factors driving climate change and variability. Historical records reveal the cyclical nature of climate phenomena, such as ice ages, warm periods, and fluctuations in atmospheric composition. Through meticulous examination of paleoclimatic archives, researchers can reconstruct ancient temperature fluctuations, precipitation levels, and atmospheric conditions, shedding light on the interplay between various environmental parameters. Furthermore, the exploration of historical climate events enables the identification of long-term trends, facilitating the discernment of natural variability from anthropogenic influences. Concurrently, studying historical climate patterns allows for a comparison with contemporary climatic trends, aiding in the evaluation of the extent to which recent changes align with or deviate from historical norms. By scrutinizing historical climate data alongside modern observations, scientists can gauge the degree of deviation in key climatic indicators, offering pivotal insights into the rapidity and magnitude of ongoing climate shifts. Additionally, this comparative analysis contributes to our understanding of the interconnectedness of global climate processes, underlining the intricate relationships between oceanic circulation patterns, atmospheric dynamics, and terrestrial feedback mechanisms. Moreover, historical climatic analysis affords a broader perspective on the potential implications of current trends, elucidating the precursors and catalysts of major climatic shifts in different geographical regions. By integrating historical context with contemporary observations, climatologists gain a robust foundation for projecting future climate trajectories, enhancing the precision of predictive models and scenarios. Furthermore, by delving into historical climate patterns and current trends, we are better equipped to comprehend the multi-faceted ramifications of climate change, enabling proactive adaptation and mitigation strategies to be formulated and implemented at local, regional, and global scales.

Methodologies for Interpreting Paleoclimatic Data

Paleoclimatology, the study of past climates, relies on a wide array of methodologies to interpret historical climate data and reconstruct ancient climatic conditions. This field integrates various disciplines, including geology, paleontology, archaeology, and climatology, to unravel the complex tapestry of Earth's climatic history. One fundamental technique is the analysis of proxy data, which provides indirect evidence of past climates. Proxy records encompass a diverse range of sources such as ice cores, tree rings, sediment layers, and fossilized pollen, each offering unique insights into past environmental conditions. These proxies serve as invaluable archives, preserving essential data over geological timescales. For instance, ice cores drilled from polar ice sheets offer detailed snapshots of ancient atmospheres, enabling scientists to examine fluctuations in greenhouse gas concentrations and atmospheric temperatures over thousands of years. Similarly, tree rings provide vital clues about historical precipitation patterns and temperature variations, while sediment cores extracted from ocean floors yield critical information about past sea surface temperatures and oceanic circulation patterns. Another pivotal methodology involves the use of isotopic analysis. Isotopes, variants of chemical elements with differing numbers of neutrons, serve as natural tracers of environmental processes. By examining stable isotopic ratios in natural materials, researchers can infer past climatic conditions. For example, the ratio of oxygen isotopes in marine microfossils can shed light on ancient ocean temperatures and ice volume, unraveling the dynamics of past glacial-interglacial cycles. Furthermore, the analysis of terrestrial and marine sediments provides valuable data on past climates. Sedimentary deposits contain a wealth of information, including mineral composition, grain size, and organic content, all of which can unveil past environmental changes. With advancements in technology, sophisticated analytical techniques, such as spectrometry and imaging, have revolutionized the study of paleoclimatic data, allowing for finer resolution in reconstructing historical climates. Moreover, the integration of statistical methods and computer modeling has enabled scientists to synthesize complex datasets and generate comprehensive reconstructions of past climates. These innovative approaches have facilitated an enhanced understanding of ancient climates and their implications for contemporary and future climate trends, empowering us to make informed decisions in addressing climate change c hallenges.

The Role of Data Science in Climatology

Data science serves as a pivotal force in revolutionizing our understanding of climatology. By harnessing the power of advanced analytics, machine learning, and big data processing, data scientists play a crucial role in deciphering complex climatic patterns and predicting future trends with unprecedented accuracy. Leveraging sophisticated algorithms and computational tools, data scientists can analyze vast volumes of historical climate data to extract valuable insights and identify recurring patterns in environmental dynamics. Furthermore, the integration of remote sensing technologies and geographic information systems enables data scientists to gather real-time atmospheric and oceanic data, enriching our understanding of contemporary climate behaviors. This intersection between data science and climatology empowers researchers to develop intricate climate models that simulate various scenarios, facilitating the assessment of potential climate change impacts and informing adaptive strategies. Moreover, data-driven approaches allow for the identification of key indicators and drivers of climatic shifts, offering valuable inputs for policymakers and stakeholders to develop targeted interventions and resilient infrastructure to mitigate climate-related risks. The integration of interdisciplinary expertise, including meteorology, ecology, and geoscience, with data science amplifies the collective capacity to address climate challenges and foster sustainable practices. As we continue to navigate the intricate web of climate complexities, the role of data science stands as an indispensable ally, paving the way for innovative solutions and informed decision-making in safeguarding the planet's environmental equilibrium.

Modeling Techniques: From Theory to Practice

In the realm of climatology, modeling techniques serve as indispensable tools for transforming theoretical constructs into practical applications. These techniques are the linchpin in our quest to decode the complex language of the Earth's climate system and its historical patterns. Embracing a multidisciplinary approach, modeling harnesses the power of data science, atmospheric physics, and computational methods to simulate and predict climate dynamics. At the heart of modeling lies the integration of vast datasets derived from a multitude of sources, including satellite observations, ground-based measurements, ice core samples, and historical records. These datasets provide the foundational bedrock upon which intricate models are built. Climatologists employ diverse modeling approaches, ranging from simplified conceptual models to sophisticated numerical simulations. Each method serves a distinct purpose, allowing researchers to probe specific facets of climate behavior. Through these models, scientists can experiment with different scenarios, evaluate the potential impact of various environmental factors, and explore the interconnectedness of complex climate variables. Whether scrutinizing temperature trends, precipitation patterns, or atmospheric circulation, modeling illuminates the underlying mechanisms driving climate variations. Furthermore, these techniques enable us to dissect the nuances of past climate events, providing crucial insights into the evolution of Earth's climate over millennia. Yet, modeling is not devoid of challenges. Uncertainties persist, stemming from the inherent complexities of climate interactions and the limitations of our current knowledge. Acknowledging these uncertainties, researchers continuously refine and validate their models, refining parameters and incorporating new data to enhance predictive accuracy. As we traverse the expanse between theory and practice, the marriage of modeling techniques with empirical observations offers a gateway to unraveling the enigmatic tapestry of Earth's climate. Ultimately, this symbiotic relationship empowers us to transcend theoretical abstractions and grasp tangible phenomena, bestowing upon us the imperative tools to forecast and comprehend the profound consequences of a changing climate.

Case Studies: The Impact of Historical Climate Events

Case studies serve as pivotal tools in understanding the tangible implications of historical climate events on various ecosystems and human societies. These empirical investigations enable us to delve into the intricate interplay between past climatic anomalies and their repercussions, offering profound insights that are paramount in shaping our present and future responses to environmental challenges. By examining specific instances where historical climate events have significantly altered landscapes and ecosystems, we can grasp the magnitude of such phenomena on a granular level.

One such case study revolves around the notorious Dust Bowl of the 1930s in the United States. This catastrophic event was triggered by prolonged drought conditions, coupled with poor land management practices, leading to severe soil erosion and agricultural devastation across the Great Plains. The Dust Bowl serves as a vivid illustration of how historical climate events can precipitate widespread ecological and socio-economic upheaval, compelling us to rethink resource management and resilience measures in the face of changing climatic patterns.

Additionally, the analysis of the Little Ice Age (between the 14th and 19th centuries) presents another compelling case study. This era of cooler temperatures had far-reaching consequences, ranging from altered agricultural practices and food scarcity to demographic shifts and cultural adaptations. Such historical incidents demonstrate the ripple effects of climate fluctuations on human civilizations and underscore the intricate relationship between climatic variations and societal dynamics.

Furthermore, the examination of past extreme weather events, such as hurricanes, floods, and heatwaves, provides valuable insights into the vulnerability of different regions to varying climatic stressors. By investigating these occurrences, we gain a nuanced understanding of the multifaceted impacts of historical climate events on infrastructure, public health, and the natural environment. These case studies serve as cautionary narratives, urging us to fortify our preparedness and mitigation strategies in anticipation of future climatic uncertainties.

In essence, delving into the impact of historical climate events through comprehensive case studies not only enriches our knowledge of past environmental transformations but also equips us with invaluable precedent-based wisdom to navigate the complexities of modern climatology. These meticulous explorations serve as cornerstones in cultivating adaptive responses and informed policies to effectively address the challenges posed by contemporary and future climate scenarios.

Predictive Accuracy: Assessing the Models

In the realm of climatology, as in any scientific discipline, the assessment of predictive accuracy plays a pivotal role in shaping the credibility and applicability of models. With the growing importance of anticipating future climatic conditions, the quest for refining the accuracy of predictive models has become paramount. This pursuit necessitates a comprehensive evaluation of various modeling approaches, input parameters, and validation techniques. At the core of assessing model accuracy lies the need to compare the simulated outcomes with real-world observations from historical and contemporary datasets. Rigorous statistical analysis is indispensable in establishing the reliability and limitations of these models. Moreover, an understanding of the inherent uncertainties and assumptions associated with these models is crucial in providing a holistic assessment. Through thorough sensitivity analyses and cross-validation procedures, scientists endeavor to quantify the predictive uncertainties and identify the potential biases within the models. Furthermore, the utilization of advanced machine learning algorithms and computational tools has revolutionized the process of model assessment, enabling researchers to disentangle complex interactions and nonlinear relationships within climatic systems. As we delve into the intricacies of predictive accuracy assessment, it becomes evident that the interdisciplinary nature of this endeavor extends beyond conventional climatological boundaries. Collaborative efforts with experts from diverse fields such as statistics, computer science, and environmental engineering are essential in enriching the methodological spectrum and fostering innovation in model evaluation. The implications of this assessment extend far beyond the realm of academia, permeating into governmental policy-making, urban planning, and infrastructure resilience strategies. By critically evaluating the predictive accuracy of climatic models, society can make informed decisions aimed at mitigating risks and adapting to the ever-evolving climate scenarios. In summary, the meticulous scrutiny of predictive accuracy not only enhances the scientific rigor of climatological research but also underpins the foundation of proactive measures for safeguarding our planet's ecological equilibrium.

Future Scenarios: Anticipating Extreme Weather Patterns

As humanity grapples with the complexities of a rapidly changing climate, the ability to anticipate extreme weather patterns becomes paramount for mitigating potential risks and enhancing resilience. By extrapolating historical climatic data and leveraging advanced modeling techniques, researchers and experts are delving into the realm of future scenarios, seeking to forecast how our planet's climate may evolve in the coming decades. Anticipating extreme weather patterns entails an intricate fusion of interdisciplinary knowledge, drawing from meteorology, climatology, and environmental sciences. It requires a meticulous analysis of past climate records coupled with the integration of cutting-edge technologies to simulate potential climatic trajectories. The identification and interpretation of key indicators such as temperature fluctuations, precipitation patterns, and atmospheric dynamics serve as foundational elements in envisioning future climate scenarios. Through the synthesis of these interconnected factors, we gain insights into the potential manifestation of extreme weather events, ranging from intense heatwaves and droughts to powerful storms and heightened precipitation. These forecasts enable societies and governing bodies to formulate adaptive strategies, bolster infrastructure, and enact policy measures aimed at safeguarding communities against the disruptive impacts of extreme weather. Moreover, the anticipation of future climate scenarios provides invaluable foresight for various sectors including agriculture, urban planning, and emergency response, allowing for proactive preparations and resource allocation. As we navigate the intricate tapestry of future climate scenarios, it becomes evident that a comprehensive understanding of the dynamic interplay between natural processes and anthropogenic influences is crucial in envisioning the trajectory of our planet's climate. With each insight gleaned from the nexus of historical context and technological innovation, we fortify our capacity to anticipate, adapt, and thrive in the face of evolving weather patterns.

Future scenarios serve as compass points, guiding us toward resilient and sustainable solutions, empowering us to confront the challenges posed by an uncertain yet compelling climatic landscape.

Policy Implications: Informing Climate Strategies

Effective climate policies require a robust understanding of historical climate data and predictive modeling to inform decision-making and mitigate potential risks. By delving into the complexities of past climatic variations, policymakers gain valuable insights into the challenges that communities may face in the future. This chapter examines the pivotal role of informed climate strategies and their policy implications in addressing and adapting to changing climate patterns.

Understanding historical climate data is essential for crafting policies that can withstand the rigors of tomorrow's climate realities. Policy formation must be rooted in a comprehensive assessment of past weather events, enabling leaders to anticipate future trends and develop adaptive measures. By leveraging advanced modeling techniques and statistical analyses, policymakers can tailor their strategies to meet the unique needs of different regions and ecosystems. Such targeted approaches are crucial for building resilience and fostering sustainable development in the face of evolving climatic conditions.

Informed climate strategies also play a vital role in advocating for international cooperation and coordination. By engaging in knowledge sharing and collaborative efforts, global policymakers can align their initiatives with shared climate goals and create a unified front against the impacts of climate change. This collaboration extends beyond national borders, encompassing multilateral agreements and concerted action plans to address the interconnected nature of climate-related challenges. Through diplomacy and informed advocacy, policy frameworks can pave the way for cohesive responses to complex environmental transitions.

Furthermore, these strategies hold immense potential for influencing economic and technological innovation. A proactive approach to climate policy encourages investment in renewable energy, sustainable infrastructure, and green technologies. By promoting a transition towards low-carbon economies, policymakers can drive forward-thinking solutions that not only combat climate change but also foster job creation and economic growth. Moreover, by incorporating climate adaptation strategies into urban planning and disaster risk reduction, policymakers can bolster societal resilience and safeguard vulnerable communities against extreme weather events and environmental hazards.

The culmination of these efforts rests on the ability of policymakers to implement evidence-based strategies backed by rigorous scientific research and interdisciplinary collaboration. In this context, policymakers must engage with scientific communities, private sectors, and local stakeholders to co-create effective, adaptable, and inclusive policies. By integrating diverse perspectives and fostering a culture of continuous learning, policymakers can propel innovative solutions that address the dynamic and interconnected challenges posed by a changing climate. The unfolding policy landscape, informed by historical insights, predictive modeling, and collaborative governance, presents an opportunity to forge a resilient, sustainable future for generations to come.

Conclusion: Synthesizing Past Insights for Future Resilience

As we culminate our exploration of the implications for climatology and the profound impact of historical climate trends on present and future environmental dynamics, it becomes increasingly evident that synthesizing past insights is paramount in fortifying the resilience of our future endeavors. The amalgamation of historical climate data, predictive modeling, and policy implications enables us to not only comprehend the ramifications of previous climate events but also empowers us to chart a course towards sustainable and resilient strategies. Through the integration of interdisciplinary approaches, from paleoclimatic analysis to cutting-edge data science methodologies, we gain an unparalleled vantage point from which to anticipate and adapt to the challenges posed by evolving climatic patterns. By delving into case studies that illustrate the tangible repercussions of historical climate variations, we are able to extract valuable lessons that can inform and refine our preparedness for extreme weather scenarios and their associated risks. Moreover, the discernment of the intricacies of predictive accuracy propels us towards refining our models and enhancing our capacity to forecast future climatic trajectories with greater precision. Such advancements not only bolster our scientific understanding but also furnish decision-makers and policymakers with invaluable insights to tailor adaptive measures and formulate resilient climate policies. Consequently, the synthesis of the wealth of knowledge derived from historical climate analyses and predictive modeling harnesses the potential to steer us towards a future characterized by heightened resilience and sustainability. It is imperative for present and future generations to acknowledge the significance of such integrated efforts in order to navigate the complexities of an ever-evolving climate landscape with foresight, agility, and fortitude. In sum, by integrating historical insights with contemporary advancements, we pave the way for a future that thrives amid the challenges of climate variability and change, poised to create a harmonious coexistence between humanity and the natural world.

Chapter 16

Unlocking the Mystery of Bird Migration

Avian Migration

Bird migration is a phenomenon that has captured the fascination of scientists and naturalists for centuries. This remarkable behavior, observed across diverse bird species, involves the regular seasonal movement of birds from one geographical location to another. The primary drivers of avian migration include factors such as resource availability, breeding requirements, and climatic conditions. As birds are highly attuned to changes in their environment, they have developed intricate strategies to navigate vast distances with remarkable precision. Understanding the fundamental aspects of bird migration entails delving into key concepts and terminologies that form the foundation of this awe-inspiring natural spectacle. Central to the study of migration is the concept of migratory connectivity, which refers to the linkages between breeding and non-breeding areas of a species. This interconnectedness underscores the significance of preserving habitats across migratory routes, emphasizing the cooperative conservation efforts needed on a global scale. Additionally, the notion of stopover sites, essential resting points for migratory birds during their long journeys, highlights the critical importance of maintaining these areas to ensure the successful completion of migration. Equally essential is comprehending the adaptive advantages of migration, aligning with concepts such as energy optimization, avoiding adverse conditions, and maximizing reproductive success. These fundamental aspects lay the groundwork for a deeper exploration of the historical, ecological, and environmental dimensions of avian migration, offering profound insights into the marvels of the natural world.

Historical Perspectives on Bird Migration

Throughout history, human societies across the globe have marveled at the phenomenon of bird migration. Ancient civilizations recorded their observations and interpretations of this natural spectacle, often imbuing it with mystical significance. In ancient Greek and Roman cultures, migratory birds were perceived as divine messengers, carrying the wisdom of the heavens on their journey. The enduring influence of such beliefs can be traced through early mythologies and religious texts.

As human understanding of the natural world advanced, so did our knowledge of bird migration. Pioneering thinkers in the fields of ornithology and natural history, such as Aristotle and Pliny the Elder, made significant contributions to the early understanding of avian movement patterns. Their writings provided valuable insights into the seasonal behaviors of various bird species and laid the groundwork for future studies.

The age of exploration brought about a new era of discovery and curiosity about bird migration. As explorers ventured into uncharted territories, they encountered unfamiliar avian species undertaking extraordinary journeys. The documentation of these encounters expanded the global perspective on bird migration, revealing the interconnectedness of ecosystems across continents. Notable explorers like Charles Darwin and Alexander von Humboldt added to this body of knowledge through their extensive observations and writings, sparking widespread interest in the migratory habits of birds.

In the 19th and 20th centuries, advances in scientific methodologies and technologies facilitated more comprehensive studies of bird migration. Ornithologists began to employ innovative techniques for tracking and observing migratory birds, leading to unprecedented revelations about their routes, seasonal movements, and physiological adaptations. Notable scientific expeditions, such as those led by John James Audubon and Roger Tory Peterson, expanded our understanding of bird migration while promoting conservation efforts.

The historical perspectives on bird migration provide a rich tapestry of insights, blending cultural perceptions with scientific inquiry. These narratives underscore the enduring fascination with avian movement and the integral role that migratory birds play in shaping our understanding of the natural world.

Modern Tracking Technologies and Techniques

Advancements in technology have revolutionized the study of bird migration, enabling scientists to delve deeper into the mysteries of avian travel. Modern tracking technologies and techniques play a pivotal role in unraveling the intricate patterns and behaviors associated with bird migration. One of the primary methods utilized is satellite telemetry, which involves fitting birds with lightweight transmitters that communicate with orbiting satellites, providing real-time data on their movements. These transmitters are designed to withstand the rigors of avian flight and are instrumental in tracking long-distance migrations across continents and oceans. Additionally, geolocators, miniaturized devices that record light levels to estimate location, have significantly contributed to our understanding of migratory pathways. The precise data obtained from these devices has allowed researchers to map out previously unknown migration routes and stopover sites. Radar technology has also proven invaluable in monitoring nocturnal migration, capturing the scale and dynamics of bird movements during the cover of darkness. Furthermore, remote sensing techniques, such as weather radars and thermal imaging, have provided critical insights into avian behavior, aiding in the identification of key staging areas and flyways. Genetic tools, including stable isotope analysis and DNA fingerprinting, have shed light on population-specific migration strategies and connectivity between breeding and wintering grounds. Integrating these various tracking methods has enhanced our comprehension of the complex mechanisms guiding avian migration. By combining data from diverse sources, researchers can construct comprehensive models of migratory systems and gain a holistic view of the ecological factors influencing these behaviors. As technology continues to evolve, new frontiers in tracking and monitoring bird migration are being explored, offering unprecedented opportunities to explore avian movements with unparalleled precision and detail.

The Role of Geographical Landmarks

Geographical landmarks play a crucial role in the complex phenomenon of bird migration. These natural features serve as navigational aids for avian species during their long and arduous journeys across continents. Mountains, coastlines, rivers, and other prominent landforms act as visual cues that guide birds along their established migration routes. The memorization of these landmarks is crucial for the survival of migrating birds, as they provide reference points that help maintain the correct direction and pace.

Mountains, with their towering presence and distinct silhouettes, serve as unmistakable indicators for avian migrants. Birds have been observed to use mountain ranges as guides, following their contours to reach specific breeding or wintering grounds. Likewise, coastlines and major bodies of water also factor prominently in avian navigation. Coastal features shape the flight paths of migratory birds, influencing their trajectory and facilitating orientation over vast stretches of open ocean.

Rivers act as critical corridors for migrating birds, offering both sustenance and geographical guidance. These waterways serve as natural highways, guiding avian travelers and providing essential resources for resting and refueling. Moreover, the strategic positioning of these landmarks creates recognizable pathways that many species rely on year after year. The familiarity of these routes allows for efficient and reliable navigation, minimizing the risks associated with the taxing journey.

Understanding the significance of geographical landmarks sheds light on the remarkable adaptability and resilience of migratory birds. It accentuates the intricate relationship between these natural features and the instinctual behavior of avian species. Furthermore, acknowledging the vital role of geographical landmarks underscores the importance of conservation efforts directed towards preserving these essential elements of the natural landscape. As such, it is imperative to recognize and protect these features to ensure the continued success of avian migration and uphold the ecological balance that supports this awe-inspiring phenomenon.

Meteorological Influences on Migration Patterns

Bird migration is a phenomenon deeply influenced by meteorological factors, with weather patterns playing a pivotal role in shaping the routes and timing of avian movements. The intricate interplay between atmospheric conditions, such as wind patterns, temperature variations, and precipitation, profoundly impacts the migratory behavior of birds. Understanding these meteorological influences is essential in unraveling the complex tapestry of avian migration patterns. Wind dynamics significantly affect bird migration, particularly during the crucial periods of departure and arrival. Birds often capitalize on favorable winds to facilitate their journeys, utilizing tailwinds to gain momentum and conserving energy. Conversely, adverse wind conditions can impede progress and necessitate alterations in flight paths. An in-depth comprehension of prevailing wind systems is therefore vital in comprehending the intricacies of avian migration. Temperature variations also exert a substantial influence on migration, serving as a key determinant in the timing of departures and arrivals. Birds are highly attuned to temperature cues, using thermal gradients to optimize their flights and minimize energetic costs. Changes in temperature can trigger crucial decisions regarding the initiation or cessation of migrations, highlighting the critical significance of meteorological conditions in shaping avian behaviors. Furthermore, precipitation patterns play a significant role in dictating stopover locations and foraging opportunities along migratory routes. Birds navigate through landscapes in response to rainfall patterns, seeking out suitable habitats and food sources that are contingent upon meteorological variables. Understanding these intricate relationships between birds and meteorological phenomena is pivotal in elucidating the underlying mechanisms driving avian migration patterns. Meteorologists, ornithologists, and conservation biologists collaborate to employ advanced techniques such as radar and satellite data to monitor and analyze how birds interact with weather systems during their journeys. These interdisciplinary efforts shed light on the nexus between meteorology and bird migration, enriching our comprehension of this captivating natural spectacle.

Genetic and Evolutionary Implications

Bird migration, a marvel of natural history, is intricately intertwined with genetic and evolutionary processes that have shaped avian species over millennia. The study of genetics offers valuable insights into the mechanisms driving migratory behavior and the evolutionary adaptations that allow birds to undertake these incredible journeys. Genetic research has revealed that migratory patterns are determined by a complex interplay of inherited traits, environmental cues, and selective pressures. Through genetic analyses, scientists have identified specific gene variants associated with migratory behavior, shedding light on the genetic basis of navigation, orientation, and the timing of migrations. Furthermore, studies of the evolutionary history of migratory species have provided a deeper understanding of how migration has evolved in response to changing environments and ecological conditions. By examining the genetic diversity and population dynamics of migratory birds, researchers can discern patterns of migration routes, stopover locations, and seasonal movements, offering crucial insights into the adaptive significance of migratory behaviors. Exploring the genetic and evolutionary implications of bird migration also unveils the intricate relationships between migratory species and their habitats, as well as the potential impacts of human-induced environmental changes on these ancient migratory pathways. This intersection of genetics, evolution, and migration not only enhances our comprehension of the natural world but also underscores the urgency of conserving critical migratory corridors and addressing the challenges posed by habitat loss, climate change, and other anthropogenic threats. As we delve deeper into the genetic and evolutionary underpinnings of avian migration, we gain a profound appreciation for the resilience and adaptability of migratory species, while recognizing the interconnectedness of all life forms within the tapestry of our planet's ecosystems.

Case Studies: Iconic Migratory Species

Throughout the annals of natural history, certain migratory species have captured the imagination of scientists, conservationists, and nature enthusiasts alike. The study of iconic migratory species provides a fascinating glimpse into the intricate mechanisms governing avian migration and the diverse challenges these species face. Among the most renowned migratory birds are the Arctic Tern, renowned for its extraordinary annual journey from the Arctic to the Antarctic and back; the awe-inspiring Peregrine Falcon, revered for its remarkable transcontinental flights; and the enigmatic Bar-tailed Godwit, which holds the record for the longest non-stop flight during migration. Each of these species presents a unique set of behavioral, physiological, and ecological adaptations that enable them to undertake their awe-inspiring migratory journeys. By delving into the specific case studies of these iconic migratory species, we gain profound insights into the adaptive strategies that have evolved over millennia to facilitate their survival. Furthermore, these case studies shed light on the critical role of stopover sites along migration routes, the influence of environmental cues on navigation, and the profound impact of global climate patterns on the timing and success of migrations. Moreover, the study of iconic migratory species serves as a powerful reminder of the interconnectedness of ecosystems, and the need for concerted conservation efforts to safeguard the habitats upon which these remarkable birds depend. As we delve into the vivid accounts of these iconic migratory species, we uncover a wealth of knowledge that not only enriches our understanding of avian migration but also underscores the urgency of protecting these marvels of the natural world in the face of ongoing environmental changes.

Impact of Climate Change on Migration Routes

Climate change has emerged as a formidable disruptor of the world's natural systems, impacting every aspect of the environment, including the intricate phenomenon of bird migration. As global temperatures rise and weather patterns become increasingly erratic, avian species are being forced to adapt to significant alterations in their traditional migration routes and timing.

The melting of polar ice caps and changing precipitation patterns directly affect the availability of crucial stopover sites for migratory birds. Many species rely on specific habitats for resting and refueling during their long journeys, and any disturbance in these areas can have severe consequences for their survival. Furthermore, altered temperature and humidity levels can lead to mismatches in the timing of food availability, which may jeopardize the successful completion of migration for many avian species.

In addition to physical changes to the landscape, climate change also influences the prevalence of diseases and parasites along migration routes. Warmer temperatures can expand the ranges of pathogens and vectors, exposing birds to novel health risks. This increased exposure to disease can weaken populations and interrupt the delicate balance of ecosystems that depend on migratory species for various ecological functions.

Furthermore, shifting climate patterns can lead to asynchronous phenology, where the timing of key life events such as nesting, hatching, and food availability becomes mismatched. This dissonance can result in reduced breeding success and decreased population resilience, ultimately posing a threat to the long-term sustainability of migratory bird species.

The impacts of climate change on migration routes extend beyond avian populations and have far-reaching ecological implications. Disruptions to food webs and natural pollination processes, as well as the potential loss of ecosystem services provided by migratory birds, can reverberate throughout entire ecosystems, affecting other wildlife and even human livelihoods.

Addressing the challenges posed by climate change on bird migration requires interdisciplinary approaches that encompass conservation biology, ecology, meteorology, and policy-making. Efforts to mitigate these impacts include the preservation and restoration of critical stopover habitats, the implementation of international conservation agreements, the development of early warning systems for migratory bird health, and the reduction of greenhouse gas emissions through sustainable practices.

Successfully protecting migratory bird species from the perils of climate change demands proactive measures at the local, regional, and global levels. The collective commitment to safeguarding the integrity of migration routes and the habitats upon which avian species depend is essential for preserving the awe-inspiring phenomenon of bird migration for generations to come.

Conservation Strategies and Challenges

Conserving migratory bird species presents a complex challenge that demands a comprehensive approach. Addressing the various threats facing avian migrants requires strategic planning, international collaboration, and innovative solutions. One of the primary challenges in conservation efforts is the vast geographical range these birds cover during their migrations. As they traverse through numerous countries, coordinating conservation actions becomes crucial to ensure their survival. Additionally, the diverse habitats utilized by migratory birds further complicate conservation endeavors, as it necessitates a multi-pronged strategy tailored to each specific ecosystem. Habitat loss due to urbanization, deforestation, and agricultural expansion profoundly impacts the well-being of migrant birds. Therefore, conserving crucial stopover sites and breeding grounds is imperative for maintaining healthy populations. Another significant threat comes from pollution, particularly the detrimental effects of pesticides and other chemical contaminants on migratory bird populations. The accumulation of these toxins along migration routes can lead to dire consequences for the birds and their ecosystems. Moreover, climate change poses a formidable challenge to the conservation of migratory birds, altering traditional migration patterns and impacting essential resources such as food and nesting sites. Mitigating these challenges demands coordinated global action, with a focus on policy development, sustainable land use practices, and public awareness campaigns. International agreements and partnerships play a critical role in addressing conservation challenges, facilitating cross-border cooperation, data sharing, and research collaborations. By fostering a unified front towards avian conservation, the international community can work together to protect critical habitats, implement sustainable land management practices, and enact legislation to safeguard migratory bird species. Embracing cutting-edge technologies, such as geolocators, satellite tracking, and remote sensing, enhances our understanding of avian migration behavior and assists in identifying priority areas for conservation. Integrating this knowledge with adaptive management strategies can aid in the formulation of effective conservation plans that address present and future challenges. By engaging local communities and stakeholders, conservation efforts can also contribute to socioeconomic development and bolster support for conservation initiatives. Furthermore, educational programs and outreach activities are vital in cultivating a deeper appreciation for migratory birds and fostering a sense of stewardship among the public. Through collaborative research, interdisciplinary approaches, and unwavering dedication, we can overcome the obstacles posed by conservation of migratory bird species and ensure the perpetuation of these awe-inspiring natural wonders.

Future Directions in Avian Migration Research

Research into avian migration has reached a pivotal juncture, opening up exciting possibilities for future exploration and discovery. As we look ahead, it is imperative to outline the key trajectories that will shape the landscape of avian migration research. One promising avenue is the integration of advanced tracking technologies with big data analytics. By harnessing the power of satellite telemetry, geolocators, and other cutting-edge tools, researchers can obtain unprecedented insights into the precise movements of migratory birds. This wealth of data can then be analyzed using sophisticated algorithms to elucidate complex migration patterns and behaviors.

Furthermore, there is a growing emphasis on interdisciplinary collaboration in avian migration research. The convergence of expertise from fields such as ornithology, ecology, genetics, and climatology holds immense potential for holistic understanding. By fostering partnerships across diverse disciplines, we can unravel the intricate connections between environmental factors, genetic predispositions, and physiological adaptations that drive avian migration.

In addition, upcoming research endeavors should strive to explore the implications of climate change on avian migration with greater depth and nuance. Understanding how shifting climates impact migration routes, breeding grounds, and stopover sites is critical for implementing effective conservation strategies. By simulating various climate change scenarios and their potential repercussions, researchers can proactively assess the adaptability of migratory species and identify vulnerable populations that require targeted conservation efforts.

Moreover, the burgeoning field of molecular biology presents an array of opportunities to delve into the genetic underpinnings of avian migration. Advancements in genomics and epigenetics offer a tantalizing prospect for unraveling the hereditary mechanisms that govern migratory behaviors. Unraveling the genetic code that underlies navigation abilities, energy utilization during migration, and seasonal timing of movements can pave the way for groundbreaking insights into the evolutionary basis of avian migration.

Finally, embracing citizen science initiatives can revolutionize the scale and scope of avian migration research. Through public engagement and participatory monitoring programs, a global network of bird enthusiasts and nature lovers can contribute valuable observations and data. This democratization of research empowers citizens to be active participants in scientific endeavors, thereby augmenting the volume of information and broadening the geographical reach of migration studies.

As we embark on this new frontier of avian migration research, these future directions stand poised to catalyze profound discoveries and transformative advancements in our comprehension of one of nature's most captivating phenomena.

Chapter 17

Revolutionizing Zoology with Bio-Coded Messages

Bio-Coded Messages in Zoology

Biological communication, a cornerstone of zoological studies, has always been an enigmatic and captivating field for scientists. However, recent advancements in technology and interdisciplinary collaborations have opened up new frontiers in our understanding of bio-coded messages in the animal kingdom. This chapter will delve into the intricate world of bio-codes and their significance within zoology, offering an exploration of both the historical foundations and contemporary breakthroughs that have revolutionized our perception of animal communication. At its core, this section aims to unravel the fundamental concepts that underpin bio-coded messages, shedding light on their role in shaping behaviors, ecological dynamics, and evolutionary pathways across diverse species. By illuminating the bio-linguistic landscape of the natural world, we endeavor to navigate the complex interplay between genetics, neurobiology, and environmental stimuli that form the basis of these bio-coded messages. Furthermore, we will embark on a journey through historical landmarks in biological communication studies, tracing the evolution of thought and methodologies that have shaped the current understanding of this fascinating realm. From ethological observations to molecular analyses, each milestone has contributed to our ever-growing comprehension of the subtle cues, signals, and patterns that define animal communication systems. Moreover, we will elucidate how the identification and decoding of bio-codes can provide invaluable insights into ecological interactions, social structures, and cognitive capacities within various taxa. As we navigate this terrain, it becomes evident that bio-coded messages are not only repositories of information but also windows into the rich tapestry of life on Earth. Ultimately, this section strives to lay the groundwork for the subsequent explorations into specific modalities of bio-communication, setting the stage for a comprehensive and insightful journey into the dynamic world of zoology.

Historical Overview of Biological Communication Studies

Biological communication studies have a rich and storied history that dates back to ancient civilizations. The study of animal communication has captivated scholars and naturalists for centuries, driving a continuous quest to unravel the mysteries of how living organisms convey messages within and across species. The roots of biological communication studies can be traced to early observations of animal behaviors and vocalizations by renowned naturalists such as Charles Darwin and Konrad Lorenz. Their groundbreaking work laid the foundation for understanding the mechanisms and complexities of communication in the natural world. Over time, advancements in technology and scientific methodologies have allowed researchers to delve deeper into the intricate web of signals, cues, and languages employed by diverse species. The expansion of ethology as a formal discipline in the 20th century further accelerated the systematic study of animal behavior and communication, fostering a more comprehensive understanding of the evolutionary and ecological significance of communicative behaviors. Pioneering studies on animal signaling, including seminal works on mate selection, territorial displays, and predator-prey interactions, have contributed to our evolving comprehension of the adaptive functions and underlying neural mechanisms that govern biological communication. Additionally, the integration of interdisciplinary approaches from fields such as neuroscience, genetics, and behavioral ecology has broadened the scope of biological communication studies, offering fresh insights into the genetic basis and evolutionary trajectories of communicative traits across taxa. Furthermore, the emergence of bioacoustics, chemical ecology, and visual signaling research has shed light on the multifaceted modalities through which animals exchange information, revealing the intricate connections between sensory perception, physiological constraints, and social dynamics. The historical trajectory of biological communication studies reflects an enduring intellectual curiosity and an unwavering commitment to deciphering the profound intricacies of communication systems in the natural world.

Decoding the Language: Techniques and Technologies

In the quest to understand the intricacies of biological communication, researchers have continually devised and refined a myriad of techniques and technologies for decoding the language of animals. At the forefront of these endeavors are advancements in bioacoustics, behavioral observation, neurobiology, and molecular genetics.

Bioacoustics plays a pivotal role in deciphering animal communication, wherein sophisticated recording equipment captures and analyzes the diverse sounds emitted by creatures in their natural habitats. From the haunting calls of whales reverberating through the ocean depths to the melodic songs of songbirds echoing through the forest canopy, each vocalization carries with it a wealth of information—information that speaks volumes about the social structures, mating rituals, and hierarchical dynamics of different species.

Complementing this auditory dimension is the nuanced art of behavioral observation, involving the patient and meticulous study of animal movements, gestures, and postures. This method unveils the silent but eloquent language of non-vocal communication, shedding light on subtle cues and signals that convey messages within and between species.

Furthermore, advances in neurobiology have provided invaluable insights into the neural mechanisms underpinning animal communication. By examining the intricate circuitry and chemical signaling within the brains of various creatures, scientists have gained profound understanding of the sensorimotor pathways governing vocalizations, as well as the complex interplay between cognition and communication.

Equally transformative has been the application of molecular genetics in decoding bio-coded messages. Through genetic analyses, researchers can unravel the genetic basis of vocalization patterns, identify specific genes associated with communication behaviors, and trace the evolutionary trajectories of communication systems across different taxa. The marriage of genetics and behavior has proven to be instrumental in elucidating the underlying genetic architecture shaping the diverse languages of the animal kingdom.

As the fields of bioacoustics, behavioral observation, neurobiology, and molecular genetics continue to evolve, so too does our capacity to unlock the rich tapestry of biological communication. From the harmonious symphonies of whale pods to the intricate duets of avian courtship, each revelation brings us closer to bridging the divide between humans and the captivating creatures with whom we share this planet.

Case Studies: Successful Decoding in Mammalian Species

Throughout the history of zoological research, decoding the intricate language of mammalian species has remained a constant challenge. However, recent advancements in bio-coding technologies have led to significant breakthroughs in understanding the communication patterns of various mammals. This section will delve into several compelling case studies that exemplify successful decoding initiatives in different mammalian species.

The first notable case study centers on the communication behaviors of bottlenose dolphins. By meticulously analyzing their vocalizations and body language, researchers have uncovered an astonishing array of communicative signals used by these intelligent marine mammals. From signature whistles that serve as individual identifiers to complex sequences of clicks and whistles that convey specific messages, the deciphering of dolphin communication has provided profound insights into their social structure and interactions.

In a similar vein, the study of elephant communication has yielded remarkable revelations. Through careful observation and acoustic analysis, scientists have identified an extensive repertoire of low-frequency rumbles emitted by elephants, each carrying nuanced meanings related to emotions, reproductive cues, and warnings of potential threats. The translation of these bio-coded messages has not only deepened our understanding of elephant societies but has also paved the way for innovative conservation strategies aimed at mitigating human-wildlife conflicts.

Furthermore, the case study on the language of primates, particularly chimpanzees and gorillas, showcases how bio-coding has elucidated the complexity of their non-verbal communications. By studying gestures, facial expressions, and vocalizations in their natural habitats, researchers have unveiled a sophisticated system of signals that facilitate social bonding, conflict resolution, and collective decision-making within primate communities. These insights have reshaped our perspectives on the cognitive capacities of our closest relatives in the animal kingdom.

In summary, these case studies underscore the transformative impact of bio-coding technologies on unraveling the enigmatic languages of mammalian species. By discerning the nuances of their bio-coded messages, we are granted access to the intricacies of their social dynamics, ecological roles, and emotional lives. The continued exploration of mammalian communication is poised to yield invaluable contributions to both zoological scholarship and wildlife conservation efforts.

Birds and Bio-Codes: Insights from Avian Communication

The study of avian communication offers a fascinating insight into the intricate world of bio-coded messages. Birds have long been admired for their melodious calls, but their communication extends far beyond mere songs. By delving into the realm of avian bio-codes, researchers have unveiled an astonishing array of signals, gestures, and vocalizations that signify complex messages within bird communities.

Avian communication encompasses various forms, from distinctive vocalizations to visual displays and body language. The diversity of bird species presents an equally diverse range of communication tactics, each serving unique purposes such as courtship, territorial defense, and warning signals. Furthermore, the development of advanced recording and analysis technologies has enabled scientists to capture and decode subtle nuances in avian communication that were previously inaccessible.

One prominent aspect of avian communication is the use of vocalizations to transmit information. Birds employ a wide repertoire of calls and songs, each with its own distinct purpose. For instance, certain bird species are known for their elaborate courtship rituals, where males showcase their vocal prowess to attract potential mates. Moreover, avian vocalizations play a crucial role in establishing territorial boundaries and alerting others to potential dangers in the environment.

In addition to vocal signals, visual displays and body language form an integral part of avian communication. Vibrant plumage, intricate dance-like movements, and specialized physical gestures all contribute to conveying essential messages within bird communities. These visual signals often serve as means of expressing dominance, attracting mates, or signaling imminent threats, thus playing a vital role in maintaining social cohesion and hierarchical structures among avian populations.

The insights gained from the study of avian communication have profound implications for various fields, including ecology, behavior, and conservation. Understanding the intricacies of bird bio-codes not only enhances our knowledge of avian social dynamics but also provides valuable tools for conservation efforts. By deciphering the language of birds, researchers can gain crucial insights into habitat preferences, population dynamics, and ecological interactions, thus aiding in the formulation of effective conservation strategies to protect avian species and their habitats.

Furthermore, the application of bio-coding techniques in avian communication studies holds promise for broader implications in the field of animal behavior and cognitive science. The sophisticated nature of avian communication systems offers a rich landscape for exploring the evolution of language, social learning, and cognitive abilities in non-human organisms. Through continued research and exploration, the study of avian bio-codes is poised to illuminate new frontiers in our understanding of animal communication and cognition.

The Impact of Bio-Coded Messages on Conservation Efforts

As the field of zoology evolves, the integration of bio-coded messages has revolutionized conservation efforts in profound ways. By deciphering the intricate language embedded within the biological realm, researchers and conservationists have gained invaluable insights that are directly shaping modern conservation strategies. Through the decoding of bio-codes, the comprehension of behavioral patterns, species interactions, and environmental adaptations has been elevated to a level previously unattainable. Understanding these bio-coded messages equips conservationists with the necessary knowledge to develop targeted and effective conservation interventions.

This intimate understanding of specific species' communication patterns has empowered conservationists to identify and address critical factors impacting the survival of vulnerable populations. For instance, the interpretation of bio-codes has shed light on the complex social structures of various species, allowing for the implementation of measures that safeguard crucial social dynamics within animal communities. Moreover, the insights gleaned from bio-coded messages have revealed subtle but impactful environmental cues that influence critical aspects of species behaviors and population dynamics, enabling conservationists to design interventions that mitigate adverse effects and promote thriving ecosystems.

Furthermore, bio-coded messages have played an instrumental role in fostering public awareness and advocacy for conservation initiatives. The ability to effectively communicate the intricacies of animal behaviors and ecological relationships to a broader audience has been pivotal in garnering support for conservation endeavors. By uncovering and translating these bio-coded narratives into compelling stories, the conservation community can foster a deeper connection between the public and the natural world, inspiring collective action and commitment to preserving biodiversity.

In addition, the application of bio-coded messages has transcended species-specific approaches to conservation, offering a holistic understanding of ecosystem dynamics. By deciphering the shared language of diverse species within an ecosystem, conservationists are better positioned to implement comprehensive strategies that address interconnected ecological challenges. This integrated approach to conservation has proven indispensable in mitigating the cascading impacts of disturbances and facilitating the restoration of balanced ecosystems. The utilization of bio-coded messages has transformed conservation from a reactive endeavor to a proactive, informed, and adaptive process that considers the nuanced intricacies of ecological systems.

Ultimately, the impact of bio-coded messages on conservation efforts extends beyond scientific domains, permeating societal, political, and ethical dimensions. The newfound comprehension of bio-codes has revitalized conversations surrounding wildlife preservation, human-wildlife coexistence, and global biodiversity. It has instigated paradigm shifts in how we perceive and champion the conservation of our natural world, laying the groundwork for sustainable stewardship and harmonious cohabitation with the Earth's diverse inhabitants.

Interspecies Communication: Theoretical Frameworks

The study of interspecies communication encompasses a rich tapestry of theoretical frameworks that serve as the foundation for understanding the complexity and diversity of interactions between different organisms. One such framework is the biosemiotic approach, which posits that living organisms engage in semiotic processes, utilizing signs and signals to convey information and meaning. This perspective recognizes the presence of intentional and meaningful communication across species boundaries, challenging traditional human-centric views of communication. Additionally, ethologists have contributed invaluable insights through the lens of behavioral ecology, emphasizing how evolutionary pressures shape the development of signaling systems and the role of communication in fostering adaptability and survival. Understanding the nuances of interspecies communication requires an exploration of cognitive ethology, which delves into the mental processes and mechanisms underlying these interactions. This framework sheds light on the cognitive capacities of different species, offering a glimpse into their ability to comprehend, process, and respond to communicative signals. Importantly, sociobiological perspectives bring attention to the social dynamics and evolutionary implications embedded within interspecies communication. By examining the ecological and social contexts in which communication occurs, researchers can unravel the intricate networks of interactions that influence behavior and cooperation among diverse species. Furthermore, integrating insights from cultural evolution theory broadens our understanding of the transmission and acquisition of communicative behaviors within and across species, highlighting the role of cultural practices in shaping communication patterns. Finally, the study of interspecies communication also intersects with the field of linguistics, inviting a comparative examination of linguistic structures and functions in human and non-human animal communication systems. Through the exploration of these theoretical frameworks, researchers can gain a comprehensive understanding of the multifaceted nature of interspecies communication, paving the way for groundbreaking advancements in zoological research, ecological conservation, and our relationship with the natural world.

Challenges and Limitations in Interpretation

Biocoded messages have revolutionized our understanding of zoological communication. However, this groundbreaking approach is not without its challenges and limitations. In the study of interspecies communication, a major obstacle lies in the complexity and diversity of bio-codes across different species. Each species has evolved unique methods of conveying information, making it difficult to establish universally applicable principles for interpretation. Compounding this challenge is the dynamic nature of biological systems, where bio-codes may evolve rapidly in response to environmental pressures or social dynamics within a population.

Another pivotal challenge arises from the limited capacity of human understanding to fully grasp the intricacies of non-human communication. While researchers continue to develop frameworks and methodologies for decoding bio-codes, there remains the inherent barrier of comprehending these messages through a purely human-centric lens. This limitation underscores the need for interdisciplinary collaboration, leveraging expertise across fields such as biology, neuroscience, anthropology, and linguistics to enrich our interpretations.

Furthermore, the inherent ambiguity of bio-codes presents a considerable obstacle in their interpretation. Unlike human languages with distinct grammatical structures and vocabulary, bio-codes often rely on subtle cues, context-dependent signals, and non-verbal communication. Deciphering these nuanced expressions requires meticulous observation and robust analytic tools, adding layers of complexity to the interpretation process.

It is essential to acknowledge the ethical considerations surrounding the interpretation of bio-coded messages. As we delve deeper into the realms of animal communication, we must tread carefully to avoid anthropomorphizing behaviors or making unwarranted assumptions about the intentions behind bio-coded messages. Respectful and responsible interpretation demands a commitment to objectivity, empathy, and cultural sensitivity toward non-human species.

Moreover, logistical limitations pose practical obstacles to the comprehensive interpretation of bio-coded messages. Field research, particularly in remote or inaccessible environments, presents challenges in gathering data and conducting long-term observations necessary for accurate interpretation. Additionally, technological limitations in sensor capabilities and data processing hinder our ability to capture and analyze complex bio-codes, further constraining our interpretation efforts.

As we confront these challenges, it becomes evident that the interpretation of bio-coded messages requires a holistic, multidisciplinary approach. By acknowledging these limitations and actively addressing them, we can refine our methodologies, cultivate a deeper understanding of zoological communication, and ultimately harness the full potential of bio-codes to illuminate the rich tapestry of interspecies interaction.

Future Directions in Zoological Research

As we stand at the precipice of a new era in zoological research, it is imperative to delineate the potential pathways that will shape the future of this field. One of the key directions for zoological research lies in the integration of advanced technologies such as bioacoustics, remote sensing, and molecular techniques. The development of non-invasive methods for studying animal behavior and communication will revolutionize our understanding of the intricate bio-codes embedded within diverse species. Moreover, interdisciplinary collaborations between zoologists, geneticists, ecologists, and computer scientists will open new vistas for unraveling the complexities of interspecies communication.The emergence of big data analytics and machine learning presents an unprecedented opportunity to dissect massive datasets, decode complex communication patterns, and predict ecological trends. Harnessing these cutting-edge tools will not only enhance our comprehension of animal languages but also enable us to foresee and mitigate conservation challenges. Furthermore, future zoological research must focus on exploring the connections between animal communication and climate change. Understanding how environmental alterations influence vocalizations, chemical signaling, and other forms of biological communication is paramount in safeguarding vulnerable s pecies.While technological advancements are crucial, there is an equally pressing need to conduct long-term field studies and behavioral observations. Equipped with modern tracking devices and monitoring systems, researchers can gain insights into migratory patterns, social structures, and reproductive strategies across different species. By amalgamating traditional fieldwork with contemporary methodologies, the trajectory of zoological research will be propelled toward a more holistic and nuanced understanding of animal communication networks.Additionally, fostering a global network of collaborative research initiatives is essential for the advancement of zoology. By leveraging international partnerships and sharing resources, researchers can access a broader spectrum of biodiversity and cultural knowledge, thereby enriching their studies and contributing to a more comprehensive understanding of animal communication.Finally, ethical considerations and animal welfare should remain at the forefront of future zoological research endeavors. As technologies continue to evolve, researchers must uphold strict ethical standards, prioritize the welfare of studied animals, and ensure the responsible usage of acquired knowledge. This commitment to ethical practice will fortify public trust in scientific pursuits and foster sustainable stewardship of natural environments.With these future directions in mind, zoological research stands poised to venture into uncharted territories, unraveling the mysteries of animal communication and driving conservation efforts. By embracing technological innovation, interdisciplinary collaboration, long-term field studies, global partnerships, and ethical paradigms, the realm of zoology will undoubtedly continue to expand and transform our perception of the natural world.

Conclusion: The Broader Implications for Science and Society

As we conclude our exploration of bio-coded messages in zoology, it is crucial to acknowledge the broader implications that this field holds for both science and society. The deciphering of bio-codes not only enhances our understanding of animal communication but also has far-reaching effects on various aspects of research, conservation, and human interaction with the natural world. One of the most significant implications lies in the potential for advancements in medical science. By unraveling the intricacies of bio-codes, scientists may gain invaluable insights into the mechanisms of diseases in both animals and humans, leading to novel approaches for treatments and prevention. This breakthrough could revolutionize healthcare and pharmaceutical industries, offering solutions to pressing health challenges. Furthermore, the profound impact of bio-coded messages extends to ecological conservation efforts. Understanding the subtle languages of different species can inform strategies for habitat preservation, species management, and biodiversity conservation. Conservationists can utilize this knowledge to create more effective protection plans, mitigate human-wildlife conflicts, and restore ecosystems. The broader implications of bio-coding also extend to education and public awareness. By popularizing the concept of bio-coded communication, we can foster a deeper appreciation for the complexity of the natural world and promote environmental stewardship. Such awareness can inspire the next generation of scientists, conservationists, and policymakers, leading to a more sustainable and harmonious coexistence with nature. Finally, the study of bio-coded messages has wide-reaching implications for the ethical treatment of animals. As we gain insight into the intricate communication systems of different species, we are prompted to reevaluate our treatment of animals and recognize their cognitive and emotional capacities. This understanding can drive changes in animal welfare laws, policies, and societal attitudes towards non-human beings. In conclusion, the exploration of bio-coded messages in zoology transcends the boundaries of traditional scientific inquiry and delves into the realms of medicine, conservation, education, and ethics. By embracing the implications of this burgeoning field, we stand to unlock remarkable potentials for positive transformation in both science and society.

Chapter 18

The Language of the Deep: Understanding the Secrets of the Ocean Depths

Oceanic Language

The exploration of oceanic languages offers a captivating glimpse into the enigmatic world of marine communication. Fundamental to this exploration is the understanding that deep-sea creatures have evolved intricate systems of communication to navigate their lightless, pressurized habitats. At the heart of this investigation lies the profound realization that life in the ocean depths is not silent, as previously assumed, but rather filled with a symphony of sounds, light signals, and chemical cues. Through the study of oceanic language, we endeavor to unravel the intricacies of how marine organisms convey information, establish relationships, and navigate the awe-inspiring abyss. To comprehend the nuances of oceanic language, we delve into the evolutionary pressures that have shaped the means of communication for diverse species inhabiting the deep sea. This entails an examination of sensory adaptations, including keen abilities to interpret bioluminescent displays, detect subtle shifts in hydrostatic pressure, and utilize olfactory and chemical signaling. Such mechanisms not only facilitate survival in a challenging environment but also enable sophisticated social interactions among marine organisms. Furthermore, the study of oceanic language expands beyond auditory and visual cues, encompassing the complexities of chemical messages exchanged between deep-sea denizens. These chemical signals serve as vital means for conveying information pertaining to territory demarcation, mating availability, and predator presence, thus showcasing the immense breadth of communication modalities that govern life in the ocean depths. As we embark on this journey of discovery, we recognize that augmenting our comprehension of oceanic language holds crucial implications for conservation efforts and the protection of marine ecosystems. By gaining insight into the multifaceted communication systems employed by deep-sea creatures, we are better equipped to preserve their delicate habitats and foster an unobtrusive coexistence. Through the prism of oceanic language, we witness nature's unparalleled capacity for adaptation, innovation, and interconnectedness, inspiring a sense of reverence for the intrinsic beauty concealed within the depths of our planet's oceans.

Historical Context of Marine Exploration

Marine exploration has a rich and intriguing history that dates back to ancient civilizations. The quest for understanding the secrets of the ocean depths has been an enduring endeavor, driven by a profound curiosity about the mysteries hidden beneath the waves. Historically, early civilizations such as the Phoenicians and Polynesians were among the first seafaring peoples who navigated vast oceanic expanses, driven by trade, exploration, and a deep connection to the sea. Their knowledge of marine currents, celestial navigation, and the behavior of marine organisms shaped the foundation of maritime exploration. In the medieval era, the voyages of discovery led by explorers like Christopher Columbus and Ferdinand Magellan opened new frontiers in oceanic exploration, expanding the known boundaries of the world and revealing the vastness of the oceans. The Age of Enlightenment further propelled marine exploration, with scientific expeditions seeking to unravel the intricacies of oceanic phenomena. Notable figures like Captain James Cook undertook ambitious voyages that provided valuable insights into the diverse ecosystems thriving beneath the surface of the seas. As technology advanced, the invention of diving apparatuses and submersibles allowed intrepid individuals to venture deeper into the ocean, unveiling a hitherto unseen world teeming with life and geological wonders. The pioneering work of underwater explorers such as Jacques Cousteau revolutionized our perception of the ocean depths, highlighting the need for conservation and sustainable stewardship. Today, marine exploration continues to captivate the imagination of scientists, researchers, and adventurers, each contributing to the ongoing narrative of discovery and understanding. The historical tapestry of marine exploration is marked by a relentless pursuit of knowledge, driven by an innate human fascination with the boundless expanse of the world's oceans.

Decoding Bioacoustic Signals

Bioacoustic signals play a pivotal role in the complex language of the deep sea. Within the ocean's vast and often unexplored depths, marine organisms communicate through an array of sound frequencies, creating an intricate tapestry of bioacoustic signals that are integral to their survival and behaviors. These signals encompass a diverse range of sounds, including clicks, whistles, trills, and pulsed calls, each serving unique communication purposes within the underwater world.

The profound significance of bioacoustics in marine environments cannot be overstated. It shapes the dynamics of marine ecosystems, influencing predator-prey interactions, mating rituals, territoriality, and even parental care. The symbiotic relationships between different species often rely on the exchange of bioacoustic signals to establish and maintain connections in the dark, expansive oceanic realm.

Deciphering bioacoustic signals presents a multifaceted challenge. Researchers deploy advanced hydrophones and acoustic recording devices to capture and analyze these elusive sounds, often facing technical obstacles due to the propagation of sound in water. Furthermore, understanding the context and meaning behind these signals requires an interdisciplinary approach, merging biology, acoustics, and behavioral ecology.

One remarkable example of bioacoustic communication is the mesmerizing songs of cetaceans such as whales and dolphins. These magnificent creatures produce elaborate vocalizations that convey intricate messages across vast distances. The study of cetacean bioacoustics not only sheds light on their social structures and group dynamics but also underscores the fundamental importance of preserving their habitats to safeguard their unique modes of communication.

In addition to cetaceans, many other marine organisms utilize bioacoustic signals for social cohesion, navigation, and foraging practices. From the rhythmic choruses of fish and invertebrates to the percussive beats of snapping shrimp, the underwater world resounds with an astonishing variety of bioacoustic expressions.

However, the harmony of this underwater symphony faces unprecedented threats from human activities, including anthropogenic noise pollution and disturbances from maritime traffic. These disturbances can disrupt vital bioacoustic signals, causing detrimental impacts on marine species and ultimately contributing to ecological imbalances in the ocean depths.

As we delve deeper into the enigmatic world of bioacoustics, it becomes evident that this field of study holds the key to unlocking the mysteries of marine life and fostering a deeper appreciation for the intricacies of oceanic communication. By comprehensively unraveling the bioacoustic lexicon of the deep sea, we gain invaluable insights into the interconnectedness of marine species and the fragile balance of their ecosystems.

The Role of Hydrothermal Vents in Ecological Communication

Deep within the ocean's abyss, where immense pressure and complete darkness dominate, an otherworldly ecosystem thrives around hydrothermal vents. These hot, mineral-rich springs on the seafloor create a haven for a myriad of extraordinary life forms, fostering a unique environment for ecological communication. The remarkable biological diversity surrounding these vents is symbolic of the innovative methods by which organisms interact and communicate in this extreme habitat.

At hydrothermal vents, the method of communication between different species takes on an added layer of complexity due to the distinct challenges presented by this environment. Organisms have adapted to utilize various sensory mechanisms for communication, such as chemical signaling, bioluminescence, and acoustic signals. These interactions form intricate networks of communication strategies that are essential for survival and reproduction in this harsh ecosystem.

One of the key aspects of ecological communication at hydrothermal vents is the interdependence between different species. Symbiotic relationships between organisms are common, with some species relying on others for essential resources or protection. The utilization of chemical and physical signals to attract partners, navigate the landscape, and defend territory underscores the vital role of communication in maintaining these symbiotic partnerships.

Furthermore, the presence of microbial life near hydrothermal vents has significant implications for ecological communication. Microbes play a crucial role in nutrient cycling and energy transfer within this ecosystem, influencing the communication dynamics between larger organisms. Understanding the interplay between microbial communities and macroscopic fauna is essential for comprehending the overall tapestry of communication in this unique environment.

The exploration of hydrothermal vent ecosystems has also led to groundbreaking technological advancements in the field of marine biology. Innovations such as deep-sea submersibles, remotely operated vehicles, and advanced imaging technology have provided researchers with unprecedented access to observe and record the complex interactions and behaviors occurring around these vents. This technological progress has illuminated the intricacies of ecological communication and enhanced our understanding of the interconnectedness of life in the depths of the ocean.

As we continue to unveil the secrets of hydrothermal vent ecosystems, the study of ecological communication in these environments will undoubtedly yield further discoveries that have the potential to inform conservation efforts and expand our knowledge of life on Earth. With ongoing research and technological innovation, the role of hydrothermal vents in ecological communication will remain a captivating area of exploration, offering profound insights into the hidden language of the ocean depths.

Bioluminescence: Nature's Language of Light

Bioluminescence, the production and emission of light by living organisms, is a captivating mode of communication that serves various functions in the enigmatic world of the deep sea. This phenomenon, driven by biochemical reactions within marine organisms, unveils a mesmerizing language of light that transcends the boundaries of conventional verbal expression. In the profound darkness of the ocean depths, bioluminescent organisms utilize this radiant form of communication for a multitude of purposes, ranging from predation and defense to courtship and camouflage.

Within the abyssal realms, numerous species harness bioluminescence as a means of luring prey or confusing potential predators. The beguiling display of light emitted by certain deep-sea fish and invertebrates not only attracts unsuspecting victims but also enables these creatures to evade detection and ambush their targets. Bioluminescence thus exemplifies a strategic tool in the perpetual struggle for survival amidst the perpetual gloom of the ocean's depths.

Furthermore, bioluminescent signals play a pivotal role in the complex mating rituals of many deep-sea organisms. Glowing displays serve as alluring beacons, enabling potential mates to locate one another in the vast expanse of darkness, fostering courtship and reproduction in an otherwise lightless environment. The eloquent dances and bursts of light orchestrated by bioluminescent species underscore the extraordinary richness of communication mechanisms present in the cryptic world beneath the waves.

Beyond its significance in ecological interactions, bioluminescence also holds profound implications for scientific inquiry and technological innovation. The study of bioluminescent organisms has inspired groundbreaking research in diverse fields, from biotechnology and medical diagnostics to environmental monitoring and genetic engineering. By unraveling the intricate biochemistry behind this radiant phenomenon, researchers continue to unlock new vistas of knowledge, harnessing the remarkable language of light for the advancement of human understanding and the betterment of society.

As humanity embarks on continued exploration and stewardship of the oceanic realm, the language of bioluminescence stands as a testament to the astounding diversity and adaptive ingenuity of life in the deep sea. Embracing the ethereal beauty and inherent utility of bioluminescence not only enriches our comprehension of marine ecosystems but also kindles a profound reverence for the enduring mysteries that shroud Earth's most enigmatic and least explored domains.

Interpreting Chemical Signaling in the Deep Sea

The deep sea is a realm of mystery and fascination, where marine organisms have adapted to survive in extreme conditions. Amidst this profound environment, chemical signaling plays a pivotal role in communication among diverse species. From pheromones emitted by fish to the chemical cues exchanged between predator and prey, these signals form an intricate language that governs life in the ocean depths. Interpreting this complex chemical messaging system is essential for unraveling the secrets of the deep sea.

Chemical signaling in the deep sea encompasses a myriad of interactions, each carrying its own nuanced meaning. For instance, chemical cues released by organisms can serve as beacons for navigation, enabling creatures to locate potential mates or feeding grounds in the abyssal darkness. At the same time, these cues also play a crucial role in defense mechanisms, allowing certain species to deter predators or warn conspecifics of impending danger. The ability to interpret these chemical signals opens a window into the behavioral and ecological dynamics of deep-sea ecosystems.

Moreover, chemical signaling extends beyond individual interactions to influence entire ecological communities. Symbiotic relationships, wherein different species coexist for mutual benefit, often involve chemical communication as a means of fostering cooperation. For example, chemosynthesis at hydrothermal vents depends on specialized bacteria that harness energy from chemicals exuded by the vent's geological structures. This process not only sustains these unique ecosystems but also underscores the profound significance of chemical signaling in maintaining the delicate balance of life in the deep sea.

Advancements in technology have enabled researchers to delve deeper into the intricacies of chemical communication in the deep sea. State-of-the-art equipment such as chemical sensors and analytical methods have empowered scientists to detect and analyze minute quantities of chemical compounds dispersed in the oceanic expanses. Furthermore, interdisciplinary collaborations between biologists, chemists, and oceanographers have led to groundbreaking discoveries regarding the diverse roles of chemical signaling in the realm of underwater communication.

Despite these notable strides, the study of chemical signaling in the deep sea poses numerous challenges. The vastness of the oceans, combined with the sheer diversity of marine organisms, presents immense hurdles in comprehensively cataloging and interpreting the multitude of chemical signals traversing through the depths. Additionally, human activities, such as pollution and climate change, introduce unprecedented disruptions to the natural chemical equilibrium of marine environments, further complicating the task of understanding these intricate communication networks.

In conclusion, the exploration of chemical signaling in the deep sea represents a frontier of scientific inquiry with profound implications for our understanding of marine ecosystems. By deciphering the language of chemical communication in this enigmatic domain, we gain insights into the interconnectedness of life beneath the waves and pave the way for informed conservation efforts and sustainable management of the ocean depths.

Symbiotic Relationships: A Dialogue Below Waves

Symbiotic relationships in the deep sea represent a captivating aspect of marine life, characterized by intricate interactions between different species thriving in an environment largely devoid of sunlight. The darkness of the abyssal zones conceals a rich tapestry of associations, where organisms engage in mutually beneficial partnerships that are essential for their survival. From bioluminescent bacteria that illuminate the bodies of fish to the fascinating rapport between deep-sea anglerfish and their symbiotic bioluminescent bacteria, the ocean depths harbor extraordinary examples of these interconnected relationships. The phenomenon of chemosynthesis, observed in hydrothermal vent ecosystems, is another powerful demonstration of symbiosis, where microbes convert chemicals released from the vents into energy, providing sustenance for a diverse array of life forms. These intricate partnerships underscore the resilience and complexity of deep-sea ecosystems, offering a window into the profound interdependence of marine life below the waves. Furthermore, these symbiotic relationships play a pivotal role in maintaining the delicate balance of the oceanic ecosystem. As we delve deeper into understanding these associations, we gain valuable insights into the functioning of complex ecological systems, highlighting the significance of conserving these fascinating underwater worlds. However, the fragile nature of these relationships also underscores the vulnerability of deep-sea ecosystems to anthropogenic disturbances, emphasizing the urgent need for responsible ocean stewardship. Advancing our comprehension of these symbiotic dialogues will not only expand our knowledge of the underwater realm but also contribute to developing sustainable strategies for preserving the biodiversity of the oceans, ensuring that these captivating narratives continue to unfold in the enigmatic world below the waves.

Technological Advances in Submarine Research

Technological advances in submarine research have revolutionized our ability to explore and understand the secrets of the ocean depths. With the development of unmanned underwater vehicles (UUVs) equipped with advanced sensors and cameras, scientists can delve into previously inaccessible areas of the deep sea, capturing high-resolution images and data without the need for human presence. These UUVs have enabled researchers to study marine life and geological formations in ways never before possible, shedding light on the true complexity and richness of the deep-sea ecosystems. Furthermore, advancements in acoustic technologies such as hydrophones and sonar systems have allowed scientists to listen to and map the soundscape of the ocean, revealing a symphony of biological, geological, and anthropogenic noise that shapes the underwater acoustic environment. The deployment of moored observatories and autonomous underwater gliders has also facilitated long-term monitoring of oceanographic processes, providing invaluable insights into the dynamics of deep-sea water masses, currents, and chemical properties. Moreover, the integration of satellite remote sensing with submarine research has expanded our understanding of large-scale oceanography, helping us track ocean circulation patterns, detect phytoplankton blooms, monitor sea surface temperatures, and identify potential areas of interest for further exploration. As we continue to advance in the field of submarine research, the development of new technologies such as underwater mapping systems, DNA sequencing tools for environmental DNA analysis, and deep-sea laboratories holds great promise for unraveling the mysteries of the ocean depths and enhancing our knowledge of marine communication and ecosystem dynamics.

Impacts of Human Activities on Marine Communication

Human activities have had a profound impact on marine communication, leading to significant disruptions in the natural language of oceanic ecosystems. One of the most pervasive and detrimental effects is the introduction of underwater noise pollution. The increasing presence of shipping vessels, industrial activities, and naval operations has significantly raised the ambient noise levels in many marine environments. This noise interference can disrupt the communication and navigation abilities of marine species, including whales, dolphins, and various fish species that rely on sound for essential life functions such as finding food, locating mates, and avoiding predators. Additionally, the construction and operation of offshore structures, such as oil rigs and wind farms, contribute to underwater noise pollution, further complicating the acoustic environment. Another concerning impact stems from the discharge of chemical pollutants and waste into the oceans. These pollutants can alter the composition of marine environments, affecting the physiology and behavior of marine organisms and potentially disrupting their chemical signaling and communication processes. As a result, the intricate balance of chemical cues in marine ecosystems becomes disrupted, impacting the ability of organisms to interact effectively. Furthermore, the extraction of resources from the ocean, such as overfishing and deep-sea mining, can lead to habitat destruction and the loss of biodiversity, which in turn can hinder natural communication systems among marine species. Climate change also poses a threat to marine communication, as rising sea temperatures, ocean acidification, and extreme weather events can lead to disruptions in traditional communication patterns and affect the distribution and abundance of marine organisms. Without proper mitigation measures, human-induced disturbances in marine communication could have far-reaching ecological consequences, challenging the resilience of marine ecosystems and their inhabitants. Recognizing these impacts is crucial in fostering sustainable practices and conservation efforts aimed at preserving the intricate language of the deep sea.

Future Prospects: Expanding Understanding of Ocean Depths

As society becomes increasingly attuned to the importance of preserving our natural ecosystems, there is a growing urgency to expand our understanding of the ocean depths. Advancements in technology are offering unprecedented opportunities to delve deeper into this unexplored realm and uncover its mysteries. One promising avenue for expanding our comprehension of oceanic communication involves the development of innovative monitoring systems that can withstand the extreme conditions found in deep-sea environments. These sophisticated devices offer the potential to capture real-time data on marine life and their communication patterns, providing insights into the complexity of interactions occurring beneath the waves. Moreover, the integration of artificial intelligence and machine learning algorithms holds significant promise for enhancing our ability to interpret and catalog the diverse array of signals emitted by deep-sea creatures. By harnessing these tools, researchers can not only identify known communication patterns, but also discover new forms of interaction previously hidden from view. Collaborative efforts among scientists, oceanographers, engineers, and conservationists are essential in driving forward these explorations. Research institutions, governmental organizations, and private enterprises need to unite in their commitment to advancing this frontier of knowledge. Education and public engagement initiatives play a pivotal role in fostering widespread awareness and investment in the exploration of oceanic communication. Encouraging the next generation of scientists and advocates to pursue careers in marine biology, ecology, and technological innovation will be critical in sustaining and accelerating these endeavors. As we look toward the future, ethical considerations must also remain at the forefront of our pursuits. The protection and respect of oceanic life forms are paramount as we seek to decode their intricate language. Ethical guidelines and international collaboration are imperative to ensure that our quest for knowledge does not come at the expense of disrupting or damaging the delicate ecosystems thriving in the ocean depths. By embracing an interdisciplinary approach and a steadfast commitment to ethical practice, we can lay the groundwork for a future where the secrets of the ocean depths are unveiled without compromising the very wonders we seek to understand.

Chapter 19

Geology's New Frontier: Unearthing Hidden Narratives

Modern Geological Exploration

Geological exploration has undergone a profound transformation over the years, evolving from traditional field-based techniques to the utilization of cutting-edge geophysical methodologies. This revolution in exploration methods has significantly enhanced our ability to comprehend the Earth's subsurface structure and composition. The shift towards modern geological exploration is marked by the integration of advanced technologies such as satellite imaging, LiDAR (Light Detection and Ranging), ground-penetrating radar, and seismic surveys. These tools enable geoscientists to acquire detailed geological data with unprecedented precision and accuracy, offering insights into previously inaccessible terrains. The evolution of exploration techniques has not only revolutionized the field of geology but has also paved the way for interdisciplinary collaborations with experts in remote sensing, data analytics, and artificial intelligence. Moreover, the synergy between traditional field observations and state-of-the-art geospatial technologies has led to a more holistic understanding of geological processes. By unraveling the intricate layers of Earth's history and dynamics, modern geological exploration plays a pivotal role in addressing contemporary challenges such as resource management, environmental conservation, and natural hazard mitigation. With the advent of advanced geophysical techniques, geologists can now delve deeper into the Earth's narrative, deciphering hidden stories preserved beneath the surface. Through this chapter, we embark on a captivating journey through time and technology, tracing the remarkable progression of geological exploration from its rudimentary origins to the forefront of innovation and scientific discovery.

Advancements in Geophysical Techniques

Geological exploration has entered a new era marked by significant advancements in geophysical techniques that are revolutionizing the way we unravel the Earth's hidden narratives. These cutting-edge methods offer unprecedented insights into the subsurface, presenting a wealth of data that was previously inaccessible. One such breakthrough is the widespread adoption of advanced imaging technologies like 3D seismic surveys, ground-penetrating radar, and magnetotellurics, which have enhanced our ability to visualize and interpret geological structures with remarkable clarity and precision. Additionally, the advent of drone-based aerial surveys and satellite remote sensing has allowed for comprehensive mapping of terrains, enabling geologists to analyze vast expanses of land from an entirely novel perspective.

Furthermore, the integration of multi-physics approaches has ushered in a new era of comprehensive data collection, consolidating information from various sources such as seismic, magnetic, gravitational, and electromagnetic surveys. This holistic view of subsurface properties facilitates a more nuanced understanding of geological features, sedimentary basins, fault lines, and hydrocarbon reservoirs. Another pivotal development lies in the refinement of borehole geophysics, where instruments deployed deep within the Earth's crust yield detailed measurements that shed light on subsurface dynamics and geological formations.

In recent years, the application of advanced computational modeling, such as finite element analysis and numerical simulations, has greatly amplified the predictive capabilities of geophysical investigations. By harnessing the power of supercomputing and algorithmic algorithms, geoscientists can simulate complex geological processes, sedimentation patterns, and tectonic movements, paving the way for enhanced risk assessment, resource evaluation, and environmental impact studies.

Moreover, the emergence of real-time monitoring systems equipped with state-of-the-art sensors and IoT devices has enabled continuous surveillance of geological phenomena, including seismic activities, volcanic eruptions, and ground deformations. This unparalleled level of monitoring not only contributes to hazard mitigation but also provides invaluable data for ongoing research and scientific endeavors.

The continuous evolution of geophysical techniques underscores the dynamic nature of geological exploration, propelling the discipline into uncharted territories of discovery and innovation. As these advancements continue to unfold, they hold the promise of unlocking a trove of hidden narratives buried beneath the Earth's surface, reshaping our understanding of the planet's intricate geological history and its profound implications for numerous facets of human civilization.

Role of Data Analysis and Machine Learning

Data analysis and machine learning have revolutionized the field of geology, serving as powerful tools for uncovering hidden narratives within the Earth's crust. By harnessing the potential of big data and sophisticated algorithms, geologists can now process immense volumes of complex geological information to extract valuable insights and make groundbreaking discoveries. This section delves into the pivotal role played by data analysis and machine learning in reshaping modern geological exploration. The integration of these computational techniques has enabled geologists to decipher intricate patterns in seismic, gravitational, and magnetic data, thereby illuminating the geological composition and history of the Earth's subsurface. Through advanced computational modeling, geoscientists can simulate and interpret various geological processes, shedding light on phenomena such as tectonic plate movements, volcanic eruptions, and sedimentation over geological time scales. Moreover, machine learning algorithms facilitate the identification of subtle correlations and anomalies within vast geological datasets, leading to the identification of potential mineral deposits, hydrocarbon reservoirs, and geological hazards. Furthermore, the utilization of machine learning in geology enables predictive analytics, allowing for the forecasting of geological events based on historical and real-time data. These technologies also aid in automating repetitive tasks such as image recognition in satellite imagery, enabling geologists to focus on higher-level analysis and interpretation. As geoscience enters the era of big data, the integration of data analysis and machine learning holds the promise of unlocking unprecedented insights into the Earth's dynamic processes and ancient histories, ultimately transforming the way we perceive and understand our planet.

Case Studies: Breakthroughs and Discoveries

In the realm of modern geology, case studies are pivotal in showcasing the practical applications of theoretical concepts and cutting-edge technologies. Throughout history, numerous groundbreaking discoveries have reshaped our understanding of geological processes, Earth's history, and even the potential impacts on future developments. These case studies not only illuminate the complexities of our planet but also serve as catalysts for further exploration and innovation. One such significant case study involves the discovery of an ancient seabed buried deep within the Earth's crust, revealing a plethora of fossilized marine life from millions of years ago. Through meticulous analysis of sedimentary layers and fossil records, researchers were able to unveil the environmental conditions and ecological dynamics of this long-lost marine ecosystem, providing profound insights into prehistoric oceanography and evolutionary biology. Another remarkable breakthrough stemmed from the utilization of advanced seismological imaging techniques to uncover an underground reservoir of precious minerals, previously hidden from conventional survey methods.

The detailed mapping of this mineral deposit not only revolutionized mining practices but also shed light on the complex geological processes that led to its formation. Furthermore, the integration of isotopic dating methods with geochemical analysis allowed scientists to reconstruct the timeline of volcanic eruptions and their influence on past climate patterns, leading to a comprehensive understanding of volcanic activity's far-reaching implications. These case studies emphasize the interdisciplinary nature of modern geological research, transcending traditional boundaries to merge geology with diverse fields such as chemistry, physics, biology, and engineering. Additionally, they underscore the indispensable role of technological advancements in driving paradigm shifts within the field, emphasizing the critical symbiosis between scientific ingenuity and the tools at our disposal. By leveraging these exceptional case studies as benchmarks, geologists continue to push the boundaries of knowledge, unveiling hidden narratives embedded in Earth's timeless tapestry and shaping the future of geological exploration.

Interpreting Geological Strata and Their Stories

The geological strata are repositories of Earth's history, each layer holding secrets of the past waiting to be deciphered. By carefully studying these formations, geologists can unlock a wealth of information about the planet's evolution, environmental conditions, and the forces that have shaped the landscape over millions of years. Interpreting geological strata requires meticulous observation, analysis, and a holistic understanding of Earth's processes.

Geological strata provide a record of environmental changes, such as shifts in climate, sea level fluctuations, and volcanic activities. By analyzing the composition and characteristics of different layers, geologists can reconstruct ancient environments and climates. For instance, the presence of certain fossils, minerals, or sediment types within specific strata can offer valuable insights into the prevailing conditions at the time of deposition.

Furthermore, understanding the sequence of geological strata enables geologists to unravel the narrative of Earth's tectonic history. Tectonic movements, including the shifting of continents, the formation of mountain ranges, and the opening and closing of ocean basins, leave discernible imprints in the layers of rock. By correlating these structural features with known geological events, researchers can piece together the dynamic story of plate tectonics and its impact on the Earth's surface.

Moreover, the study of geological strata plays a crucial role in resource exploration and environmental assessment. By interpreting the age and characteristics of rock layers, geologists can identify potential mineral deposits, hydrocarbon reservoirs, and groundwater sources. Additionally, an understanding of the geological past aids in predicting geological hazards, such as landslides, earthquakes, and volcanic eruptions, contributing to better land use planning and hazard mitigation strategies.

The interpretation of geological strata is not just a static exercise in reading rock layers—it is a gateway to understanding the dynamic interplay between Earth's processes, climate variations, and the evolution of life. It is a testament to the interconnectedness of geological, biological, and environmental systems which have molded the world we inhabit today. As technology and analytical methods continue to advance, the stories embedded in the geological strata will continue to unfold, providing profound insights into the history and future of our planet.

Linking Geological Features to Historical Climate Changes

The study of geological features offers a unique perspective on unraveling the historical narrative of Earth's climate. By analyzing various geological formations such as sedimentary layers, fossils, and glacial deposits, researchers can piece together a detailed chronicle of past climatic conditions. Through this process, it becomes evident that geological strata are not just silent repositories of ancient history but interactive storytellers that seamlessly weave together the intricate tapestry of Earth's climatic evolution. Sedimentary rocks, for instance, provide valuable clues about the environmental conditions prevalent during their formation. Their composition, structure, and organic content can offer insights into temperature fluctuations, sea level changes, and the prevalence of different ecosystems across geological time periods. Moreover, the presence of certain fossils within specific strata serves as a testament to the climatic conditions that facilitated the existence and extinction of diverse life forms. Additionally, the study of isotopic ratios in minerals and organic materials enables scientists to reconstruct past temperature variations, thereby shedding light on the fluctuations in global climate. Furthermore, the analysis of glacial landforms and moraines offers compelling evidence of ancient ice ages, highlighting the dynamic shifts in Earth's climate over millions of years. These revelations not only deepen our understanding of historical climate change but also hold profound implications for addressing contemporary climate challenges. By deciphering the interconnected language of geological features and historical climate changes, we gain crucial insights into the earth's sensitivity to external influences, natural climate variability, and the long-term impact of human-induced alterations. This integrative approach not only enhances our ability to predict future climate trajectories with greater accuracy but also underscores the urgency of sustainable practices to mitigate adverse climate outcomes.

Collaborative Efforts Across Disciplines

In the realm of contemporary geology, collaborative efforts across disciplines play a pivotal role in advancing our understanding of the Earth's geological narratives. Geology is inherently multidisciplinary, and harnessing the expertise of various scientific realms such as paleoclimatology, geochemistry, geophysics, and even archaeology can lead to groundbreaking insights. By integrating knowledge from diverse fields, researchers can gain a more comprehensive perspective on geological phenomena and unravel hidden narratives within the Earth's strata. Collaborative efforts also bridge gaps between traditional and emerging disciplines, fostering innovation and dynamic approaches to geological exploration.Geoarcha eologists work hand in hand with geologists to decode ancient landscapes and human histories, uncovering how past civilizations interacted with their geological environments. This collaboration helps in piecing together the puzzle of human existence and its interconnectedness with geological processes. Additionally, the amalgamation of geological and climatological expertise enables scholars to reconstruct ancient climates and map out the evolution of Earth's environment over millennia. Moreover, collaborations with experts in technology and data science facilitate the development of advanced computational models for analyzing geological data, leading to more accurate interpretations and predictions. The interplay between different disciplines enriches research methodologies, as geneticists, biologists, and ecologists contribute their knowledge to understand the deep-time interactions between geological changes and the evolution of life on Earth. Furthermore, the fusion of environmental science and anthropology sheds light on the intricate relationship between human societies and geological landscapes, offering valuable insights into sustainable resource management and societal resilience. It is through these interdisciplinary connections that the geological community can address complex challenges such as natural hazard mitigation, resource exploration, and environmental conservation with greater depth and breadth. In essence, collaborative efforts across disciplines not only elevate the scientific exploration of the Earth's geological history but also pave the way for holistic and innovative solutions to real-world issues.

Challenges and Limitations in Contemporary Geology

Geological exploration has made remarkable strides in recent years, but it is not without its challenges and limitations. One of the key challenges faced by contemporary geologists is the complexity and sheer scale of geological systems. The Earth's geological processes operate on vast timescales and are influenced by a myriad of interconnected factors, making it difficult to fully comprehend and predict changes. Additionally, accessing certain geologically significant areas, such as remote or inhospitable regions, presents logistical and safety challenges for researchers. Another significant limitation is the need for advanced technologies and skilled personnel to interpret and analyze the immense volumes of geological data generated from modern exploration methods. This reliance on technology can pose barriers for researchers working in resource-constrained environments or lacking access to state-of-the-art equipment and software. Furthermore, geological research often requires interdisciplinary collaboration, which can be slowed down or impeded by differing methodologies, language barriers, or conflicting priorities among experts from various fields. In addition, as we delve deeper into our understanding of Earth's geological phenomena, ethical considerations about environmental impact, conservation, and indigenous rights become increasingly complex. Balancing the pursuit of knowledge with responsible stewardship of natural resources poses a profound challenge for contemporary geologists. Another critical consideration is the unpredictable nature of geological events such as earthquakes, volcanic eruptions, and landslides. These natural hazards can pose significant risks to both researchers and local communities, adding an element of uncertainty to geological exploration. Moreover, in the realm of contemporary geology, there exists a continual need for funding and resources, as cutting-edge equipment and large-scale expeditions require substantial financial investment. Securing sustained support for geological research amid competing scientific priorities presents its own set of obstacles. Despite these challenges and limitations, the field of contemporary geology holds tremendous promise and potential. By openly addressing and actively seeking solutions to these difficulties, geologists can pave the way for groundbreaking discoveries and a more comprehensive understanding of Earth's geological history and future.

Future Prospects: Next-Gen Geological Technologies

The continuing advancement of technology is poised to revolutionize the field of geology, promising a future full of exciting prospects and opportunities. Next-generation geological technologies are anticipated to not only address current limitations but also open up new frontiers for exploration and understanding of the Earth's history. One such area of development lies in the use of advanced remote sensing techniques, including satellite imaging and drone technology, which allow for comprehensive and high-resolution mapping of geological features over wide areas, providing invaluable insight into the Earth's subsurface structures. These cutting-edge tools offer a means to penetrate previously inaccessible terrains and unravel the hidden narratives locked within the Earth's crust. Additionally, the integration of artificial intelligence (AI) and machine learning algorithms holds immense potential in enhancing the analysis and interpretation of geological data, enabling geologists to extract meaningful patterns and correlations from vast datasets with unprecedented efficiency. Furthermore, the utilization of advanced geological modeling software facilitates the simulation and visualization of complex geological processes, offering a virtual platform for testing hypotheses and refining predictive models. The convergence of these technologies is expected to significantly streamline the process of prospecting, exploration, and resource management, ushering in an era of more informed decision-making in the geological industry. Moreover, the development of novel sensor technologies, such as advanced seismological instruments and terrestrial laser scanning systems, promises to provide greater accuracy and precision in mapping geological structures, enabling a deeper understanding of fault lines, tectonic movements, and subsurface composition. This enhanced understanding not only contributes to better hazard assessment and disaster preparedness but also offers a window into the geological history of our planet. The ongoing innovations in the field of geology are not only revolutionizing traditional practices but are also paving the way for interdisciplinary collaborations with fields such as climatology, ecology, and archaeology, facilitating a holistic comprehension of Earth's intricate systems and their dynamic interactions. As next-gen geological technologies continue to evolve, these advancements will undoubtedly shape the future of geological exploration, uncovering hidden narratives and shedding light on the Earth's enigmatic past.

Summary and Implications for Future Research

As we conclude our exploration of next-generation geological technologies, it becomes evident that the future of geological research holds promise and potential for groundbreaking discoveries. The integration of cutting-edge technologies such as advanced remote sensing, artificial intelligence, and big data analytics presents unprecedented opportunities for understanding Earth's dynamic processes. These tools enable geoscientists to collect and interpret vast amounts of data, facilitating detailed analysis and modeling of geological phenomena with exceptional precision and accuracy. This shift towards data-driven approaches not only enhances our predictive capabilities but also allows for a deeper understanding of the intricate interplay between Earth's various systems. As a result, the implications for future research in geology are both profound and multifaceted. One of the key implications lies in the realm of climate change and its impact on geological landscapes. By leveraging these next-gen technologies, researchers can delve deeper into unraveling the complex relationship between geological processes and climate dynamics, shedding light on how past climatic events have shaped the Earth's surface and informing our understanding of potential future changes. Furthermore, the use of advanced geological tools opens new avenues for studying natural hazards and mitigating associated risks. Through real-time monitoring and analysis, researchers can proactively identify geological anomalies and anticipate potential environmental threats, contributing to improved disaster preparedness and response strategies. Additionally, the integration of interdisciplinary approaches harnesses the collective expertise of geologists, environmental scientists, engineers, and data analysts, fostering innovative collaborations that push the boundaries of geological research. This convergence of diverse perspectives not only offers holistic insights into Earth's complex systems but also paves the way for holistic solutions that address pressing societal and environmental challenges. As we embrace these advancements, it is crucial to acknowledge the ethical and sustainability dimensions of deploying next-gen geological technologies. Responsible and transparent usage of these tools is imperative to ensure that the advancements serve the common good and uphold environmental stewardship. Moreover, training and educating the next generation of geoscientists in the proficient use of these technologies will be pivotal in driving sustainable and impactful research initiatives. Looking ahead, the horizon of geological research appears illuminated with endless possibilities, beckoning us to embark on a transformative journey towards unlocking the Earth's hidden narratives and shaping a more resilient and informed future.

Chapter 20

Uncovering the Secrets of Ancient Civilizations

Civilizational Archaeo-Cryptography

Throughout history, ancient cultures have utilized architecturally encoded languages for a multitude of reasons, blending both practicality and ritualistic purpose. The built environment became a canvas for the expression of ideology, power dynamics, and spiritual beliefs. Within the confines of these monumental structures lie hidden messages waiting to be deciphered, offering invaluable insights into the cultural, political, and religious landscapes of bygone eras. By delving into the intricate details of architectural configurations, researchers can unravel the cryptic language embedded within these edifices, shedding light on the enigmatic practices and cosmologies of antiquity. From the precision of alignment in pyramids and temples to the symbology etched into their facades, every facet speaks to a sophisticated form of non-verbal communication, hinting at the existence of profound meanings waiting to be uncovered. These architectural languages not only served as practical guides for construction but also functioned as mnemonic devices, communal markers, and spiritual conduits. As such, they constitute an indispensable resource for comprehending the intricate fabric of these ancient societies. Through this exploration, we begin to comprehend the universal human tendencies towards symbolism and the integration of such linguistic elements into the very structures that shaped the identity and consciousness of these civilizations. Understanding how ancient civilizations used architecturally encoded languages offers invaluable insights into the intellectual, artistic, and spiritual dimensions of humanity's collective past, fostering a deeper appreciation for the sophistication and richness of our historical tapestry.

Decoding Architectural Languages: Foundational Structures and Symbols

A civilization's architecture serves as an enduring testament to its cultural identity and societal norms. Each pillar, column, or arch is imbued with meaning, a silent language waiting to be deciphered. From the grandeur of ancient temples to the simplicity of nomadic dwellings, every structure tells a story. In order to interpret these narratives, archaeologists and historians employ a multi-disciplinary approach, drawing on expertise in art history, anthropology, and architectural studies.

The foundational structures of a civilization's architecture often reveal its core values and beliefs. The layout of a city, the design of public buildings, and the arrangement of private residences all speak to the social hierarchy and community organization. For instance, the construction of monumental public spaces may signify the importance of communal gatherings and collective identity. Similarly, the placement and size of religious edifices can shed light on the spiritual significance attributed to different deities or belief systems within a society.

Moreover, the symbols and motifs carved into architectural elements offer rich insights into the culture's cosmology and mythos. Whether it's the intricate carvings adorning a temple façade or the geometric patterns found in domestic structures, these decorative features are not merely ornamental; they are part of a visual lexicon that conveys narratives of creation, heroism, and the relationship between humanity and the divine. Through meticulous analysis, researchers can unravel this symbolic vocabulary, piecing together the meanings encoded within every frieze, relief, or fresco.

As technology continues to advance, new tools have emerged that enable more detailed exploration of architectural languages. Cutting-edge imaging techniques and digital modeling afford unparalleled opportunities to reconstruct and analyze historical structures with unprecedented accuracy. This digital frontier allows scholars to virtually navigate ancient buildings, identify patterns in design, and simulate the play of light and shadow on architectural elements, thus deepening our understanding of the messages embedded in ancient constructions.

In sum, the decoding of architectural languages represents a compelling endeavor that offers profound insights into the cultural, religious, and social dimensions of past civilizations. By scrutinizing foundational structures and symbols, we gain a deeper appreciation for the intricate ways in which our ancestors communicated their values, aspirations, and worldview through the enduring medium of architecture.

The Role of Artifacts in Communicating Cultural Norms

Artifacts serve as invaluable conduits of cultural norms, offering profound insights into the beliefs, values, and daily lives of ancient civilizations. These tangible remnants of bygone eras hold within them the essence of societal customs, traditions, and practices, acting as visual and material representations of a culture's collective identity. By scrutinizing these artifacts, archaeologists can decipher the intricate web of customs that governed the lives of our predecessors. From pottery and tools to adornments and ceremonial objects, each artifact weaves a narrative that unveils the social structure, religious inclinations, and artistic expressions of ancient societies.

The significance of artifacts transcends their physical forms; they encapsulate the spirit of an era, providing windows into the socio-cultural landscapes that flourished centuries ago. Whether it's a meticulously crafted piece of jewelry symbolizing status and hierarchy or a utilitarian object reflecting the ingenuity of ancient craftsmanship, each artifact mirrors the aspirations, fears, and values of its creators. The meticulous study of these relics allows us to untangle the intricate tapestry of traditions and rituals that once permeated everyday life.

Furthermore, artifacts also offer critical insights into the interactions between different civilizations, shedding light on trade networks, cross-cultural influences, and the diffusion of ideas across vast geographical expanses. Through extensive analysis of these objects, scholars gain a deeper understanding of how diverse societies coexisted, adapted, and forged connections with one another. From the silk roads of antiquity to maritime trade routes, artifacts stand as testimonies to the cultural interplay that shaped our global history.

Beyond their historical and anthropological significance, artifacts play a pivotal role in contemporary society by fostering a sense of connection and continuity with our ancestors. Museums and exhibitions showcasing these relics provide a platform for modern audiences to engage with the rich tapestries of human heritage, cultivating empathy and understanding for the myriad ways in which societies have evolved over time. Ultimately, the study of artifacts enables us to bridge the chasm between the past and the present, ensuring that the enduring legacies of ancient civilizations continue to enrich and inspire future generations in their quest for knowledge and enlightenment.

Script and Scripting: Unraveling Ancient Texts

The study of ancient scripts and inscriptions holds the key to understanding the rich tapestry of human history. Across various civilizations, written language has been a powerful tool for recording knowledge, beliefs, and societal structures. As we delve into deciphering ancient texts, we embark on a journey through time, seeking to uncover the mysteries that lie beneath the surface.

Scripts from civilizations such as the Egyptian, Maya, Mesopotamian, and Indus Valley offer glimpses into their respective cultures and belief systems. These intricate forms of writing were often closely intertwined with religious practices, governance, and trade, representing a sophisticated means of communication that mirrored the complexities of the societies they served.

To unlock the intricate messages encoded in these ancient scripts, modern archaeologists and linguists employ a myriad of tools and techniques. From traditional methods like epigraphy and palaeography to cutting-edge technologies including multispectral imaging, 3D scanning, and computational linguistics, the arsenal for script decryption continues to expand. Such advancements enable us to discern faded or obscured inscriptions, identify recurring patterns, and compare scripts across geographical regions and historical periods, thus deepening our comprehension of ancient languages and their evolution over time.

The decipherment process itself is akin to solving a captivating puzzle, necessitating a blend of linguistic expertise, historical context, and creativity. It demands a keen eye for detail, as well as an acute awareness of cultural nuances and symbolic representations. Moreover, it often requires collaboration among experts from diverse fields, fostering interdisciplinary dialogues that enrich our understanding of the past.

Beyond merely transcribing ancient texts, the act of decoding scripts allows us to resurrect forgotten voices and unfold narratives that have endured millennia of silence. It opens windows into the way thoughts were expressed, laws were codified, and myths were perpetuated, offering profound insights into the intellectual, religious, and artistic dimensions of bygone civilizations.

In essence, the endeavor to unravel ancient texts represents an intricate dance between the contemporary and the ancient, weaving together threads of empirical analysis and speculative imagination. Through this process, we unshroud the eloquence of ancient languages, their intrinsic beauty, and the wisdom that resonates across epochs.

Technological Tools for Archeological Insights

Archaeology, as a field, has witnessed extraordinary advancements in recent decades, greatly enhancing our ability to decipher the secrets of ancient civilizations. Technological tools have revolutionized the way we approach archaeological investigations, offering unparalleled insights into historical societies and cultures. One of the most significant technological developments in archaeology is remote sensing technology, which enables researchers to conduct non-invasive surveys of archaeological sites. This technology includes aerial photography, LiDAR (Light Detection and Ranging), multispectral imaging, and ground-penetrating radar, allowing archaeologists to identify subsurface features without extensive excavations. The utilization of drones equipped with advanced imaging systems has also proven to be an invaluable tool for surveying and mapping expansive areas with precision and efficiency. Furthermore, Geographic Information Systems (GIS) has emerged as a crucial component in archaeological research, enabling scholars to analyze spatial data, create detailed maps, and model past landscapes and settlement patterns. In addition, the application of 3D scanning and printing technology has revolutionized artifact analysis and preservation. By creating precise digital replicas of archaeological finds, researchers can study objects in fine detail while preserving the originals. Moreover, advances in DNA and isotope analysis have offered remarkable insights into ancient populations, migration patterns, and dietary habits, providing a more comprehensive understanding of past human societies. Cutting-edge laboratory techniques, such as mass spectrometry and next-generation sequencing, have expanded our ability to extract and analyze genetic material from archaeological specimens, shedding light on historical demographics and connections between ancient populations. Additionally, advancements in spectroscopy and chemical analysis have facilitated the identification and characterization of archaeological materials, including pottery, metals, and pigments, aiding in the reconstruction of ancient trade networks and manufacturing processes. As technology continues to evolve, the integration of artificial intelligence and machine learning algorithms is poised to transform archaeological data analysis, enabling the rapid processing of vast amounts of information and the identification of complex patterns within archaeological datasets. These technological innovations collectively empower archaeologists to gain deeper insights into the intricacies of ancient civilizations and unlock the mysteries of our collective human heritage.

Interpreting Socio-Political Hierarchies through Excavation Discoveries

Excavation discoveries play an indispensable role in unraveling the intricate tapestry of socio-political hierarchies that characterized ancient civilizations. By meticulously unearthing artifacts, architectural remains, and material culture, archaeologists gain invaluable insights into the power structures, governance systems, and social dynamics of past societies. Through a careful analysis of spatial layouts, evidence of significant infrastructure, and the distribution of resources, researchers can discern the stratification within communities and the extent of centralized authority. The implications of these findings extend far beyond mere historical curiosity, shedding light on issues of power distribution, societal organization, and the mechanisms of governance. Moreover, the examination of ritual sites, monuments, and tombs often reveals symbolic associations with leadership, religious ideologies, and the prestige of ruling elites. Such excavations offer a unique window into the ritualistic, ceremonial, and commemorative practices that underpinned political authority and social cohesion. Furthermore, the investigation of urban planning, fortifications, and defensive structures yields critical insights into the protection of ruling centers and the display of might against external threats, providing a glimpse into the geopolitical landscape and territorial control. The study of burial customs, grave goods, and mortuary practices allows for a nuanced understanding of social differentiation, status symbols, and the role of ancestors and lineage in perpetuating hierarchical orders. Additionally, the deciphering of inscriptions, seals, and administrative documents provides evidence of bureaucratic apparatus, taxation systems, and the codification of laws, highlighting the mechanisms through which authority was exercised and disputes were mediated. In essence, the meticulous analysis of excavation discoveries enables us to construct a vivid tapestry of ancient socio-political hierarchies, illuminating the complexities of governance, social stratification, and the interplay between authority and community in civilizations of bygone eras.

Religious Symbolism and Its Influence on Community Cohesion

Religious symbolism has played a paramount role in shaping the social fabric of ancient civilizations. The deciphering of religious symbols embedded in archaeological artifacts holds the key to understanding the dynamics of communal life and the power structures within these societies. These symbols were not only expressions of faith but also served as unifying motifs that reinforced the collective identity of a community.

The study of religious symbolism provides invaluable insights into the belief systems, mythologies, and rituals practiced by our ancestors. It unveils the intricate web of meanings interwoven with societal norms and behaviors, shedding light on the fundamental values that governed daily life. By unraveling the significances behind these symbols, researchers can glean a deeper understanding of the spiritual, moral, and ethical codes that bound these ancient societies together.

Furthermore, through the analysis of religious iconography, we can discern the hierarchical structures and power dynamics prevalent within these communities. The depiction of deities, religious leaders, and sacred rituals on art and artifacts offers clues about the distribution of authority and the roles of individuals within the societal hierarchy. This exploration of symbolism enables us to grasp the mechanisms through which religious institutions influenced governance and social organization.

Moreover, the study of religious symbolism highlights the interconnectedness of various cultural groups and regions through trade and migration. The presence of similar motifs and deities across different civilizations suggests shared belief systems and cultural exchanges, fostering unity amidst diversity. It underscores the capacity of religious symbols to transcend linguistic and geographical barriers, fostering cohesion among disparate communities and civilizations.

In essence, the analysis of religious symbolism elevates our comprehension of ancient societies, offering profound insights into the ideological underpinnings, social structures, and interconnectedness of diverse cultures. These findings not only enrich our historical knowledge but also resonate with contemporary reflections on the influence of shared beliefs in fostering communal harmony and societal cohesion.

Trade, Exchange, and Economic Systems: A Linguistic Analysis

Trade, exchange, and economic systems form a vital part of ancient civilizations, reflecting their interconnectedness with neighboring societies and regions. Examining the linguistic aspects of trade and exchange provides valuable insights into the cultural, social, and economic dynamics at play during an era. The language of trade encompasses not only the literal exchange of goods and services but also the unspoken agreements, customs, and systems of reciprocity that underpin these transactions. In antiquity, the movement of goods went hand in hand with the transmission of languages, ideas, and practices, leading to the evolution of linguistic diversity and cross-cultural interactions. By studying the linguistic traces left behind in archaeological records, researchers can unravel the interconnected web of trade networks, uncovering the far-reaching impact of economic activities on language diffusion and evolution. This linguistic analysis enables scholars to map out ancient trade routes, identify key trading hubs, and analyze the linguistic exchanges that took place within these commercial centers. Moreover, understanding the linguistic nuances of trade terms, commercial contracts, and mercantile practices offers valuable clues about the economic structures, class divisions, and relational dynamics within ancient societies. It sheds light on the power differentials, negotiation strategies, and market trends prevalent in early economies. Additionally, the linguistic examination of economic systems delves into the role of languages as markers of social status, financial transactions, and resource distribution. This in-depth analysis unveils how linguistic diversity intersected with economic hierarchies, shaping the identities and interactions of diverse groups engaged in commerce. Furthermore, by scrutinizing the language of trade inscriptions, currency denominations, and trade-related correspondence, researchers gain unparalleled access to the multifaceted nature of ancient economies. This insight allows for the reconstruction of commercial partnerships, the identification of market regulations, and the interpretation of economic policies enacted by ancient authorities. Together, this linguistic scrutiny paints a comprehensive picture of the economic fabric of past civilizations, revealing the intricate connections between language, trade, and societal development.

Preservation Challenges and Solutions in Archaeology

In the field of archaeology, the preservation of ancient artifacts and sites is of paramount importance. Numerous challenges arise when attempting to safeguard these fragile remnants of history, ranging from environmental factors to human impact. One of the primary challenges is the degradation caused by natural elements such as erosion, humidity, and temperature fluctuations. These can erode delicate structures and decay organic materials, making it essential to develop innovative preservation techniques. Additionally, the encroachment of urbanization and industrial development poses a significant threat to archaeological sites, necessitating effective strategies for their conservation. Furthermore, the looting of artifacts for the illegal antiquities market remains a pressing issue, demanding increased vigilance and security measures to protect valuable cultural heritage.

To address these challenges, archaeologists have been employing a variety of solutions. Advanced technologies such as LiDAR (Light Detection and Ranging) and ground-penetrating radar have revolutionized the identification and protection of buried archaeological features. Preservation efforts also involve strategic collaborations with local communities and governments to implement legal frameworks and regulations for the safeguarding of important sites. Moreover, the establishment of museums and cultural centers not only serves as a means of public education but also provides a secure environment for the display and conservation of excavated artifacts. In addition, the development of non-invasive survey techniques and protective coatings has enabled more effective conservation of delicate material remnants.

Multidisciplinary research and international cooperation are crucial components in the ongoing battle to preserve our archaeological heritage. By sharing knowledge and best practices across borders, the global community can work together to confront the myriad challenges presented by the rapidly changing modern world. Ultimately, through sustained collective efforts, we can ensure that the invaluable insights offered by ancient civilizations endure for generations to come.

Synthesis: Integrating Data Across Multiple Civilizations

The synthesis of data across multiple civilizations represents a pivotal phase in the study of ancient cultures. By integrating findings from various archaeological sites and historical records, researchers gain a more comprehensive understanding of human civilization's diversity and interconnectedness. This process requires a meticulous approach, honed skills in pattern recognition, and a discerning eye for significant cultural markers that transcend geographical boundaries and time periods.

The synthesis of archaeological data necessitates a thorough review of artifacts, inscriptions, architectural remnants, and other relevant sources to map out commonalities and distinctions among different civilizations. It involves deciphering languages, symbols, trade routes, and technological advancements to identify patterns of influence and exchange. Through this interpretation of shared practices and materials, scholars can trace networks of interaction, cultural diffusion, and mutual inspiration across ancient societies.

Further, the integration of data enables researchers to propose hypotheses and theories about how civilizations may have interacted, collaborated, or driven conflicts. For instance, by cross-referencing findings from Mesopotamian, Egyptian, and Indus Valley civilizations, historians can infer potential trading partnerships, religious influences, or geopolitical rivalries, shedding light on the dynamics of ancient international relations.

Moreover, synthesizing data elucidates the evolution of human societies and their responses to environmental, economic, and sociopolitical challenges. By examining agricultural practices, urban planning, and governance structures, archaeologists can delineate distinct societal adaptations and innovations. This comparative approach not only deepens our knowledge of individual civilizations but also unveils broader historical trends and systemic transformations that characterize human development as a whole.

Finally, the synthesis of data across multiple civilizations serves as a foundation for fostering interdisciplinary collaborations and holistic perspectives. By incorporating insights from fields such as anthropology, sociology, linguistics, and climatology, scholars can create a more nuanced and interconnected narrative of human history. This approach transcends the confines of singular disciplines, inviting diverse expertise to converge toward a multifaceted understanding of ancient civilizations' complexities and contributions to the collective human experience.

In conclusion, the synthesis of data across multiple civilizations stands as an essential endeavor in comprehending the intricacies of ancient societies. It invites us to discern the shared narratives and distinct trajectories of human civilization, prompting further inquiry and discoveries that enrich our appreciation of the timeless legacies left behind by our forebears.

Chapter 21

Unveiling the Mysteries of Extraterrestrial Communication

Extraterrestrial Communication

The quest to understand and establish communication with extraterrestrial intelligence has fascinated humanity for centuries. Beyond the realm of speculative science fiction, this pursuit holds immense significance for our species' future. At its core, the concept of extraterrestrial communication touches upon fundamental questions about our place in the universe and the potential for interstellar cooperation and discovery. By exploring introductory concepts in this arena, we can gain a deeper appreciation for the importance of extraterrestrial communication for humanity. The foundational premise lies in the recognition that we are not alone in the cosmos, prompting our quest to decipher potential interstellar languages. This endeavor represents a profound leap into the unknown, offering the tantalizing prospect of unraveling the thoughts and knowledge of civilizations beyond our own. As we delve into this subject, it becomes clear that the study of extraterrestrial communication is intricately linked to broader scientific, philosophical, and even existential inquiries. Understanding and interpreting signals from distant stars and galaxies could provide unprecedented insights into the nature of intelligence, consciousness, and the fabric of reality itself. Moreover, by establishing contact with other intelligences, humanity stands to benefit from possible advancements in technology, science, and culture, fostering a richer tapestry of shared knowledge and experiences. The implications of successful extraterrestrial communication extend beyond scientific curiosity, carrying profound moral, ethical, and diplomatic dimensions. As we venture into the cosmos, seeking to reach out and comprehend the language of others, we confront our own perspectives and paradigms about life, existence, and the interconnectivity of cosmic entities. In essence, unlocking the mysteries of extraterrestrial communication serves as an intellectual, evolutionary, and spiritual journey, propelling us towards a more profound understanding of our place in the grand tapestry of the universe.

Theoretical Foundations of Interstellar Languages

Communication with extraterrestrial civilizations has captivated the human imagination for centuries, prompting considerable scientific and philosophical inquiry into the theoretical basis of interstellar languages. At its core, this field encompasses a multidisciplinary approach that integrates principles from linguistics, semiotics, information theory, mathematics, and astrophysics to hypothesize and decode potential modes of communication beyond our terrestrial boundaries.

Central to the quest for understanding interstellar languages is the exploration of universal concepts and mathematical principles that may serve as common semantic denominators across different species and civilizations. Theoretical frameworks such as the Universal Language of Mathematics postulate that fundamental mathematical concepts, such as prime numbers and geometric properties, could be universally comprehensible due to their intrinsic nature, potentially serving as a foundation for communication with extraterrestrial intelligence.

Furthermore, the study of semiotics and linguistic universals offers valuable insights into the possible structures and symmetries that may underpin interstellar communication. By identifying recurrent patterns in natural languages and cultural symbol systems on Earth, researchers can extrapolate potential universal structures that might extend beyond our planet, providing a framework for interpreting and generating meaningful intercultural dialogue.

In parallel, information theory plays a pivotal role in delineating the boundaries of interstellar communication by examining the efficiency, redundancy, and entropy of encoded messages transmitted from distant cosmic civilizations. Theoretical investigations often consider the optimization of information density while mitigating noise and signal degradation over vast interstellar distances, thereby elucidating the constraints and possibilities inherent in deciphering extraterrestrial transmissions.

An additional aspect of the theoretical foundations involves the potential modalities and mediums through which interstellar languages might be conveyed. Speculative examinations encompass a diverse array of possibilities, ranging from electromagnetic signals and laser-based communications to more exotic conjectures involving quantum entanglement and subatomic phenomena, reflecting the breadth of theoretical paradigms that underpin the search for interstellar linguistic constructs.

Ultimately, exploring the theoretical foundations of interstellar languages provides an intellectually stimulating and daunting endeavor, pushing the boundaries of interdisciplinary collaboration and intellectual rigor as humanity seeks to unravel the enigmatic nature of communicating with civilizations beyond the confines of our home planet.

Historical Context and Early Research

Throughout history, the idea of extraterrestrial communication has captured the imagination of philosophers, scientists, and fiction writers alike.

From ancient mythology to modern science fiction, humanity's fascination with the possibility of intelligent life beyond Earth has been omnipresent. However, it was not until the 20th century that serious efforts were made to search for evidence of extraterrestrial intelligence. The concept of communicating with beings from other worlds gained renewed attention after the first successful human space missions in the mid-20th century. The space race between the United States and the Soviet Union spurred an era of technological advancement that allowed for more comprehensive attempts to listen for potential extraterrestrial signals. In 1960, astronomer Frank Drake conducted the groundbreaking Project Ozma, using a radio telescope to listen for signals from two nearby Sun-like stars. This marked the beginning of the systematic scientific search for extraterrestrial intelligent life, also known as SETI. Throughout the following decades, numerous initiatives emerged, aiming to detect potential signals that could suggest the existence of alien civilizations. The famous Arecibo message, transmitted from the Arecibo Observatory in Puerto Rico in 1974, was one such early endeavor, providing a pictorial representation of information about human civilization to be beamed into space. Over time, the focus of such efforts shifted from merely transmitting messages to actively scanning the cosmos for incoming signals. As technology improved, the search for extraterrestrial intelligence became increasingly sophisticated, with advanced equipment and methodologies being developed to differentiate potential alien transmissions from natural cosmic phenomenon. The early research in this field laid the groundwork for future exploration and paved the way for current endeavors in the quest to decipher any potential extraterrestrial messages. These historical milestones continue to shape the ongoing pursuit of understanding and potentially communicating with intelligent life forms beyond our planet.

Analyzing Signals: From Radio Waves to Quantum Messaging

The process of analyzing signals from potential extraterrestrial civilizations represents a confluence of scientific disciplines, each contributing to our understanding of cosmic communication. At the forefront of this endeavor is the study of various forms of electromagnetic radiation, encompassing radio waves, microwaves, and even more exotic phenomena. Historically, the search for extraterrestrial intelligence began with the exploration of radio frequency signals, as these could potentially transcend the vast distances between stars and galaxies. Researchers meticulously analyze these signals, searching for any patterns or anomalies that may indicate an intelligent origin. As technology has advanced, scientists have expanded their search to include other forms of electromagnetic radiation, including optical signals and potentially even quantum communications. The burgeoning field of quantum messaging presents both incredible potential and daunting challenges. Quantum entanglement holds promise as a means of instantaneous communication over vast interstellar distances, but harnessing this phenomenon for practical communication remains a complex and elusive goal. The development of new technologies in space-based communication systems is also central to the endeavor of deciphering potential extraterrestrial messages. Advancements in satellite technology, deep-space communication networks, and advanced signal processing algorithms have significantly enhanced humanity's ability to detect and interpret faint and enigmatic signals from distant reaches of the cosmos. Furthermore, the integration of artificial intelligence and machine learning into signal analysis processes is revolutionizing our capacity to identify and classify potential extraterrestrial communications. These algorithmic tools not only sift through vast quantities of data with unprecedented precision but also adapt and learn from new information, refining their ability to distinguish meaningful transmissions from background noise. Collectively, these multifaceted approaches to analyzing signals underscore the interdisciplinary nature of the quest for extraterrestrial communication. In summary, the analysis of signals from potential extraterrestrial sources encompasses a rich tapestry of scientific inquiry, technological innovation, and speculative anticipation, all converging in the timeless pursuit of answering the fundamental question: Are we alone in the universe?

Technological Advancements in Space Communication

The quest for communication beyond our planet has driven significant technological advancements in space communication. As humanity's reach extends to the cosmos, our ability to send and receive signals over vast distances continues to evolve. Technological breakthroughs have revolutionized space communication, enabling us to explore new frontiers in interstellar dialogue. One of the key advancements is the development of high-powered radio telescopes and arrays capable of detecting faint radio signals from distant galaxies. These state-of-the-art instruments enhance our ability to listen for potential extraterrestrial transmissions, thereby expanding the scope of our search for intelligent life in the universe. Moreover, advancements in laser communications have enabled the transmission of data at unprecedented speeds, offering a promising avenue for interstellar messaging. The use of optical communication systems holds potential for establishing rapid and secure links with future space missions and potentially alien civilizations. Furthermore, the integration of quantum technologies into space communication represents a cutting-edge approach to secure and efficient interstellar messaging. Quantum communication protocols promise unparalleled security and robustness, laying the groundwork for secure exchanges with potential extraterrestrial beings. Additionally, advancements in artificial intelligence (AI) and machine learning play a pivotal role in enhancing the processing and interpretation of complex non-terrestrial messages. AI algorithms can analyze and decipher intricate patterns and signals, unlocking the potential to understand alien languages or forms of communication. Space communication technology continues to push the boundaries of what is achievable, fuelling our aspirations to establish meaningful connections with other civilizations in the universe. As we strive to unravel the mysteries of extraterrestrial communication, these advancements pave the way for an exciting era of interstellar discovery and dialogue.

Cryptographic Methods in Deciphering Non-Terrestrial Messages

Cryptographic methods play a pivotal role in the pursuit of deciphering non-terrestrial messages, where the need for robust encryption and decryption techniques is paramount. In the quest to unravel the mysteries of alien languages, cryptographic approaches serve as the bedrock for securing and interpreting extraterrestrial communications. Human civilization has long been engaged in the study of codes and ciphers, and this expertise provides a foundation for addressing the challenges inherent in decoding alien transmissions. The application of cryptographic principles involves a multifaceted approach, encompassing both classic and modern cryptographic algorithms, as well as the integration of interdisciplinary knowledge from mathematics, computer science, and linguistics. This chapter delves into the intricate world of cryptographic methods used in the endeavor to decode non-terrestrial messages. One fundamental aspect of cryptographic analysis involves the systematic examination of frequency distributions within extraterrestrial signals, akin to traditional cryptanalysis techniques applied to human languages. Moreover, the development of specialized cryptographic protocols tailored to extraterrestrial syntax enables the establishment of secure communication channels while optimizing the accuracy of language translation. Furthermore, the utilization of quantum cryptography presents an avant-garde frontier in the realm of interstellar communication, offering unprecedented levels of security and encryption resilience required for engaging with potential extraterrestrial civilizations. Quantum key distribution and entanglement-based communication protocols hold great promise in ensuring the privacy and integrity of interstellar dialogues while mitigating the risk of hostile interception or misinterpretation. As we advance towards the horizon of intergalactic interaction, cryptographic methodologies will continue to evolve in tandem with technological progress, paving the way for humanity to navigate the complexities of non-terrestrial linguistic systems. Through innovative cryptographic frameworks, we endeavor to establish a common ground for meaningful discourse with extraterrestrial beings, fostering mutual understanding and cooperation across the cosmos.

The Role of Artificial Intelligence in Interpreting Alien Languages

As humanity delves deeper into the realm of interstellar communication, the crucial role of artificial intelligence (AI) in deciphering alien languages becomes increasingly evident. In the quest to understand extraterrestrial messages, AI stands as a beacon of hope, offering unparalleled capabilities in processing and interpreting complex patterns and signals that may elude human comprehension. Leveraging advanced algorithms and machine learning, AI has the potential to bridge the communication gap between distinct civilizations, unlocking the enigmatic linguistic codes embedded within extraterrestrial transmissions. The application of AI technologies in this profound endeavor has far-reaching implications, revolutionizing our understanding of the cosmos and redefining the boundaries of interspecies communication. By harnessing AI's computational prowess, researchers can accelerate the analysis of vast datasets containing extraterrestrial communications, enabling swift recognition of recurring patterns, semantic structures, and encoding schemes, which might otherwise lay concealed within the intricate fabric of alien languages. Moreover, AI empowers scientists to discern nuanced linguistic nuances and contextual subtleties that lie at the heart of extraterrestrial discourse, shedding light on the cultural, philosophical, and scientific narratives embedded within these enigmatic transmissions. Through its adeptness at pattern recognition and probabilistic reasoning, AI is poised to undertake the Herculean task of identifying coherent symbols, syntax, and semantics within the cryptic manifestations of non-terrestrial languages, paving the way for an intimate understanding of the cultural lexicons and cosmological philosophies of otherworldly societies. However, as AI continues to spearhead the quest for extraterrestrial linguistic comprehension, ethical considerations and philosophical repercussions loom large. The prospect of entrusting AI with the monumental responsibility of mediating interspecies communication prompts introspection on the nature of cultural representation, intellectual autonomy, and moral agency in the context of intergalactic discourse. Additionally, the integration of AI and linguistic analysis necessitates vigilance regarding potential biases, preconceived assumptions, and epistemological limitations inherent in the process of interpreting and translating alien dialects. Overcoming these substantial ethical and methodological challenges requires a holistic approach that amalgamates rigorous scientific inquiry, moral deliberation, and interdisciplinary collaboration, fostering a balanced, inclusive, and principled framework for engaging with extraterrestrial intelligence through the lens of AI. Amidst these complexities, the role of AI in interpreting alien languages embodies a remarkable intersection of scientific innovation, ethical contemplation, and existential discovery, affording profound insights into the nature of intelligence, communication, and consciousness beyond the confines of Earth.

Case Studies: Notable Extraterrestrial Signals and Their Interpretations

In the quest to unravel the mysteries of extraterrestrial communication, there have been several groundbreaking case studies that have captivated the scientific community and sparked intriguing debates. One of the most notable signals was received by the Arecibo Observatory in 1977, known as the 'Wow! signal'. This enigmatic burst of radio waves lasted for 72 seconds and displayed a unique frequency characteristic unlike any natural phenomenon known to humanity. Despite numerous attempts to detect a repeat signal, the origin and nature of the 'Wow! signal' continue to baffle researchers and remains a compelling subject of study. Another remarkable case study involves the pulsar map transmitted into space from the Arecibo Observatory in 1974. This deliberate interstellar communication aimed to showcase our understanding of fundamental mathematical principles and human technology, providing potential extraterrestrial recipients with insights into our civilization. The intricate pattern contained information about our solar system, DNA structure, and even the telescope used to send the message. Furthermore, the mysterious fast radio bursts (FRBs) detected in distant galaxies have presented complex challenges in interpretation. While some hypothesize that these intense, millisecond-long signals may originate from advanced alien civilizations, others attribute them to celestial events such as neutron star mergers or black hole activity. These puzzling occurrences continue to stimulate vigorous scrutiny and imaginative theorizing. Additionally, the perplexing case of the Tabby's Star, also known as KIC 8462852, garnered widespread attention due to its erratic and irregular dimming patterns, sparking fervent speculation about potential alien megastructures. Despite various hypotheses ranging from alien megastructures to natural phenomena, the enigma of Tabby's Star serves as a reminder of the enigmatic nature of cosmic anomalies. These intriguing case studies underscore the complexity and profundity of interpreting extraterrestrial signals, igniting fervent curiosity and prompting innovative approaches to decode the language of the cosmos.

Challenges in Establishing Meaningful Dialogue with Other Civilizations

Establishing meaningful dialogue with other civilizations presents a myriad of complex challenges, stemming from not only linguistic differences but also fundamental disparities in cultural customs, societal structures, and cerebral capacities. One of the primary obstacles is the stark contrast in communication modalities and the lack of a universally shared cognitive framework. While efforts are made to decipher and decode extraterrestrial signals, the inherent bias of human perception can hinder the accurate interpretation of alien languages and symbols. Moreover, the absence of shared reference points or shared experiences further complicates the ability to bridge the gap and establish mutual understanding. Another challenge lies in the ethical considerations surrounding the act of initiating contact with potentially advanced civilizations. The implications of interstellar communication on our own society, as well as the possible consequences for the recipients of Earth's transmissions, must be approached with utmost caution and foresight. Additionally, the prospect of misinterpretation or unintended offense in cross-cultural interactions poses a substantial risk, emphasizing the critical importance of developing robust protocols and channels for intergalactic discourse. Furthermore, the very concept of 'meaning' may differ vastly across cosmic civilizations, leading to potential misunderstandings and conflicts arising from diverging interpretations. As such, it becomes imperative to engage in a multidisciplinary approach that encompasses fields such as sociology, anthropology, linguistics, and even philosophy to navigate these intricate intricacies. Lastly, the limited scope of current technological capabilities presents a significant barrier, as the vast distances and time scales involved in interstellar communication demand groundbreaking innovations in propulsion systems and signal transmission methods. Overcoming these multifaceted challenges requires an integrated effort of the global scientific community and necessitates a steadfast dedication to fostering open-mindedness, empathy, and intellectual curiosity in our pursuit of meaningful dialogues with other civilizations.

Future Prospects and Next Steps in Intergalactic Diplomacy

The quest to establish meaningful communication and diplomacy with extraterrestrial civilizations presents unprecedented challenges and opportunities for humanity. As we confront the complexities of deciphering and understanding non-terrestrial languages, we must also contemplate the potential implications of intergalactic diplomacy. The future prospects in this realm are both awe-inspiring and daunting, requiring a balanced approach that encompasses scientific rigor, ethical considerations, and global cooperation. In envisioning the next steps for intergalactic diplomacy, several key areas come into focus. First and foremost, fostering international collaboration and coordination is essential. Given the potential magnitude of such an endeavor, it is imperative for nations around the world to unite in their efforts to develop frameworks for engaging with extraterrestrial civilizations. This collaboration should extend beyond traditional geopolitical boundaries, transcending differences in ideology, culture, and societal structures. Moreover, the establishment of universal principles and protocols for interstellar communication and diplomacy is crucial. Such guidelines would serve as a roadmap for conducting respectful and mutually beneficial interactions with other intelligent beings. Furthermore, as we advance in our capabilities to interpret alien languages and signals, investing in advanced technological platforms will be paramount. Innovations in artificial intelligence, machine learning, and quantum computing hold great promise in enhancing our ability to decipher and process complex interstellar communications. Additionally, interdisciplinary research that draws from fields such as xenolinguistics, astrolinguistics, and exosemiotics will be instrumental in broadening our understanding of diverse forms of extraterrestrial communication. This necessitates sustained financial investment in scientific exploration and discovery, encouraging breakthroughs in our comprehension of non-terrestrial languages. Ethical considerations are equally vital when contemplating the future of intergalactic diplomacy. It is incumbent upon us to approach these potential interactions with respect, humility, and empathy, mindful of the profound impact they may have on both terrestrial and extraterrestrial societies. We must grapple with fundamental questions concerning the rights and autonomy of other civilizations, particularly in the context of shared resources, knowledge exchange, and potential cultural influence. In essence, the future of intergalactic diplomacy calls for a delicate balance between scientific innovation, global collaboration, and ethical conscientiousness.

The responsible pursuit of meaningful communication with extraterrestrial entities demands a holistic and multidisciplinary framework, guided by the principles of exploration, empathy, and unity.

Chapter 22

Embracing the Language of the Universe

The Quest for Universal Understanding

Throughout the annals of human history, there has been an inherent curiosity and relentless pursuit to comprehend the intricacies of universal communication. From ancient civilizations gazing at the stars to modern endeavors seeking extraterrestrial transmissions, the quest for unraveling the primary modes of cosmic dialogue has been a defining feature of our species' intellectual evolution. This chapter embarks on a profound exploration of this enduring human endeavor, delving into the profound insights garnered from past attempts and the paradigm-shifting advancements that have brought us ever closer to realizing the dream of understanding universal connectivity. It is an expedition that transcends time and space, capturing the imagination of visionaries and scientists alike and compelling them to probe the fabric of the cosmos in pursuit of knowledge that holds transformative potential for our civilization and beyond.

Decoding the Common Threads of Communication

As we delve into the realms of interstellar communication, it becomes paramount to identify the commonalities that may exist across diverse forms of universal languages. The search for common threads of communication extends beyond linguistic analysis and ventures into the intricate web of symbols, patterns, and universal constants that resonate with cosmic significance.

At the core of decoding these common threads is the fundamental realization that all intelligent civilizations, terrestrial or extraterrestrial, are likely to employ similar methods of establishing meaningful connections. This implies an exploration of not just linguistic structures but also the underlying principles of logic, mathematics, and natural laws that transcend planetary boundaries.

Efforts to decipher the common threads of communication involve interdisciplinary collaboration, drawing insights from fields such as astrolinguistics, semiotics, cognitive science, and information theory. By examining the collective wisdom of these disciplines, we gain a clearer perspective on how universal languages might manifest and evolve across the cosmos.

One of the intriguing facets in the quest to decode common threads lies in identifying recurring motifs and archetypes in different cultures and civilizations. Whether embedded in ancient mythologies or contemporary digital communications, these shared symbols serve as conduits for expressing universal concepts that bridge the gaps of spatial and temporal distances.

Furthermore, the study of fractal patterns, geometric symmetries, and harmonic frequencies holds promise in unraveling the interconnectedness of universal messages. These transcendent patterns possibly form the basis for a unified cosmic language that permeates the fabric of space, conveying sophisticated information through elegance and efficiency.

In scrutinizing the historical records of humanity's encounters with unexplained phenomena and potential extraterrestrial contacts, researchers discern tantalizing hints of overlapping communication modalities. This retrospective lens offers valuable insights into the transmission methods utilized by sentient beings and fosters a deeper understanding of the contextual cues embedded within their transmissions.

The pursuit of common threads of communication also grapples with the concept of universal grammar—a framework that seeks to delineate the universal principles governing communication systems, irrespective of their origin. It invites contemplation on whether there exists a set of inherent rules that guide the construction and interpretation of messages across disparate civilizations.

In sum, the endeavor to decode common threads of communication embodies a multifaceted exploration that transcends the constraints of individual dialects and channels the collective ingenuity of humanity toward fostering a more inclusive dialogue with the broader cosmos.

Scientific Advances in Universal Language Recognition

In the quest to understand and interpret potential universal languages, scientists and researchers have made significant strides in the field of communication beyond Earth. The development of advanced algorithms and computational models has enabled the analysis of complex patterns and signals from distant corners of the cosmos. Through collaborative efforts between linguists, astrophysicists, and computer scientists, cutting-edge technologies have been harnessed to process and decipher enigmatic messages that may hold the key to unlocking the universal language.

One groundbreaking area of research involves spectroscopy and the study of electromagnetic radiation emitted by celestial bodies. By meticulously analyzing the behavior of light and radiation from various celestial sources, scientists have identified anomalous patterns and sequences that defy conventional explanations. These anomalous signals present tantalizing possibilities for extraterrestrial communication, potentially pointing towards a form of universal language that transcends cultural and planetary barriers.

Furthermore, the advancement of artificial intelligence has revolutionized the capacity to recognize and interpret alien linguistic patterns. AI-driven systems are adept at recognizing linguistic structures and syntax, even when presented with non-terrestrial scripts or transmissions. By leveraging machine learning and neural networks, scientists are paving the way for a deeper understanding of potential universal linguistic frameworks, which could reshape our perception of communication on a cosmic scale.

Moreover, collaborative global initiatives have led to the establishment of comprehensive databases housing diverse linguistic datasets from across the cosmos. These repositories serve as invaluable resources for comparative linguistic analysis, enabling researchers to identify recurrent semantic motifs and syntactic elements that may underpin a universal language. The meticulous curation of this wealth of linguistic data empowers scientists to discern underlying patterns and conceptual constructs that could transcend the boundaries of any single world or civilization.

As humanity continues its exploration of the cosmic landscape, endeavors such as the Search for Extraterrestrial Intelligence (SETI) pursue novel methodologies to probe the depths of space for elusive linguistic phenomena. Innovative sensor arrays and telescopic systems are designed to capture and analyze faint, enigmatic signals originating from distant star systems and pulsars. The amalgamation of astrophysical insights, linguistic analyses, and technological prowess is propelling humankind towards the precipice of breakthroughs in universal language recognition and interpretation.

Ultimately, the convergence of scientific inquiry, interdisciplinary collaboration, and technological innovation heralds an era of unprecedented progress in unraveling the enigma of universal communication. As the frontiers of human knowledge expand to encompass the boundless expanse of the universe, the pursuit of understanding universal languages stands as a testament to the enduring curiosity and ingenuity of the human spirit.

Philosophical Perspectives on Universal Connectivity

As humanity makes strides in deciphering universal languages and preparing to venture deeper into the cosmos, it becomes essential to explore the philosophical implications of our evolving understanding of universal connectivity. At the heart of this exploration lies the fundamental question of our place in the universe and our relationship with other potential intelligences. Philosophers have pondered for centuries about the nature of communication, language, and interconnectedness, and now, as we stand at the threshold of interstellar interaction, these age-old inquiries take on unprecedented significance. The concept of universal connectivity prompts contemplation of our responsibilities and ethical obligations towards extraterrestrial life forms, if they exist. Furthermore, it challenges us to reevaluate our perceptions of intelligence, consciousness, and existence, pushing the boundaries of traditional philosophical discourse. By delving into these profound questions, we can navigate the complexities of imminent encounters and interactions beyond our home planet. This examination also necessitates contemplating the role of culture, belief systems, and the human condition in shaping our approach to universal connectivity. How do our cultural perspectives influence the way we perceive and interpret potential messages from extraterrestrial sources? Moreover, how might encounters with alien civilizations redefine our own understanding of identity and purpose? These are not only scientific or technological queries, but existential quandaries that demand thoughtful reflection and consideration. Additionally, it is crucial to explore the impact of universal connectivity on our sense of belonging and interconnectedness within the broader cosmic framework. As we embark on the journey of deciphering the language of the universe and seeking communion with other celestial entities, we must confront the intricate web of interconnectedness that binds all living beings. What might the recognition of a universally shared communication system signify for our understanding of interconnectedness and our place in the grand tapestry of existence? Ultimately, delving into the philosophical dimensions of this pursuit empowers us to approach the convergence of Earth's knowledge with cosmic insights with profound wisdom and thoughtful introspection.

Technological Innovations Facilitating Deep Space Communication

The quest for deep space communication has driven significant technological innovations in recent decades. This endeavor is marked by the development of advanced communication systems, capable of transmitting and receiving signals across vast cosmic distances. At the forefront of these innovations are cutting-edge antenna arrays, designed to capture and process faint signals from distant celestial bodies. These sophisticated arrays incorporate adaptive beamforming and signal processing techniques to enhance the clarity and reliability of received data, overcoming the challenges posed by cosmic noise and interference. Furthermore, advancements in quantum communication technologies show promise in establishing secure and instantaneous communication channels that could revolutionize our interactions with extraterrestrial entities.

In parallel, the proliferation of deep space probes and satellites has greatly expanded our capacity to gather and disseminate information across the universe. Equipped with state-of-the-art transceivers and data processing units, these autonomous explorers serve as vital conduits for transmitting earthly knowledge into the depths of outer space. The integration of artificial intelligence and machine learning algorithms within these probes has exponentially enhanced their capability to interpret and respond to non-terrestrial communications, amplifying our potential for meaningful interstellar dialogue.

Moreover, the advent of optical communication systems has opened new frontiers in deep space communication by leveraging photons to transmit information at unprecedented speeds. Laser-based communication technologies complement traditional radio frequency systems, offering a complementary approach for high-bandwidth data transmission over astronomical distances. These innovative solutions hold immense potential in facilitating real-time exchanges with distant civilizations while minimizing latency and signal degradation at cosmic scales.

Another pivotal breakthrough lies in the establishment of standardized protocols and communication frameworks for interplanetary and interstellar interactions. Collaborative initiatives among leading space agencies and scientific institutions have yielded robust communication standards, ensuring seamless interoperability and exchange of information between terrestrial and cosmic entities. These frameworks encompass universal encoding schemes and language structures, enabling the establishment of a common lexicon for cross-species communication, transcending traditional linguistic barriers.

The confluence of these technological marvels represents an evolutionary leap in humanity's pursuit of understanding and engaging with the cosmos. With each innovation, we edge closer to establishing meaningful connections with the enigmatic voices of the universe, opening doors to unprecedented insights and discoveries that transcend the boundaries of our home planet. Through the amalgamation of cutting-edge science and visionary tenacity, we are poised to embark on an era of truly intergalactic communication and cooperation, brought to fruition by the relentless march of technological progress.

The Role of Artificial Intelligence in Deciphering Non-Terrestrial Languages

Artificial Intelligence (AI) has emerged as a pivotal tool in unraveling the complexities of non-terrestrial languages, providing a bridge between humanity and the enigmatic messages from beyond our world. Through sophisticated algorithms and machine learning, AI systems have been instrumental in analyzing patterns, frequencies, and semantic structures embedded within extraterrestrial communications. These AI technologies enable researchers to process vast volumes of data that would be otherwise insurmountable for human analysis alone. By identifying recurring linguistic features and contextual nuances, AI facilitates the systematic breakdown of alien syntax, grammar, and vocabulary, shedding light on the underlying principles governing these obscure forms of expression. Moreover, AI-powered language translation models have been calibrated to interpret and document extraterrestrial dialects, contributing to the construction of universal lexicons that transcend terrestrial linguistic boundaries. This groundbreaking fusion of AI and interstellar linguistics extends far beyond mere comprehension; it catalyzes the development of communication protocols for potential future encounters with extraterrestrial intelligences. Furthermore, AI offers the capability to generate synthetic simulations of hypothetical non-terrestrial dialogues, enabling researchers to anticipate probable linguistic scenarios and refine strategies for establishing meaningful intergalactic communication. Ethical considerations surrounding the application of AI in deciphering non-terrestrial languages are paramount. As AI evolves, ethical frameworks must address concerns regarding privacy, consent, and representation of extraterrestrial cultures within the context of language analysis. The responsible and accountable use of AI demands meticulous oversight to ensure that interpretations and translations of non-terrestrial messages remain ethically sound, respecting the dignity and autonomy of any originating extraterrestrial societies. Amid the burgeoning potential of AI in this domain, it is imperative to cultivate interdisciplinary collaboration among linguists, ethicists, cognitive scientists, and AI specialists to navigate the intricate moral landscapes inherent to alien language studies. Consequently, the integration of AI into the expedition of comprehending non-terrestrial languages signifies a momentous leap forward in humanity's endeavor to embrace the cosmic dialogue and cultivate harmonious relations with celestial beings.

Ethical Considerations in Universal Language Studies

As humanity ventures into the realms of interstellar communication and seeks to decode non-terrestrial languages, it becomes imperative to address the ethical implications associated with such endeavors. The pursuit of understanding otherworldly languages raises profound questions regarding the potential impact on extraterrestrial civilizations. Ethicists, scientists, and policymakers face the daunting task of formulating guidelines and frameworks that govern our interactions with potential cosmic neighbors. One of the primary ethical considerations revolves around the concept of respect for alien cultures and their right to privacy. As we strive to decipher non-terrestrial languages, we must approach this endeavor with sensitivity and thoughtfulness, recognizing that our actions may have far-reaching consequences. Furthermore, there is a need to consider the potential risks inherent in initiating communication with extraterrestrial entities. Ethicists emphasize the importance of adopting precautionary measures to prevent unintended harm or misunderstanding. The responsibility to act as stewards of Earth's knowledge and wisdom also extends to our interactions on a universal scale. Alongside the ethical dimensions of communication, there are critical considerations related to the impact of our discoveries on terrestrial societies. The revelation of contact or the discovery of messages from extraterrestrial sources could have profound implications for human civilization. As such, ethicists stress the necessity of preparing society for the potential societal, cultural, and philosophical shifts that could arise from conclusive evidence of cosmic connectivity. Additionally, there is a moral imperative to ensure that any advancements in universal language studies are not exploited for nefarious purposes. Ethical frameworks must be established to guide the responsible and beneficial application of this knowledge for the betterment of all beings. Moreover, the implications of universal language studies extend beyond scientific and technological realms to encompass philosophical and existential considerations. Delving into the intricacies of non-terrestrial languages prompts contemplation of our place in the universe and challenges fundamental beliefs about consciousness, intelligence, and existence. This necessitates ethical reflection on how such knowledge may influence our perceptions of identity, purpose, and spirituality. Thus, as we navigate the uncharted waters of universal language studies, we must remain deeply cognizant of the ethical responsibilities that accompany our quest for cosmic comprehension.

Future Prospects: Integrating Earth's Knowledge with Cosmic Insights

As humanity continues to expand its understanding of the universe, the integration of Earth's knowledge with cosmic insights presents a profound opportunity for scientific advancement and philosophical enlightenment. Through the convergence of terrestrial wisdom and celestial revelations, our civilization stands on the precipice of unlocking the secrets of the cosmos. This prospect holds the potential to revolutionize not only our technological capabilities but also our fundamental understanding of existence itself. By bridging the learnings from our planet with the enigmatic knowledge garnered from the far reaches of space, we pave the path towards a more cohesive comprehension of the universe. The synergy between Earth's accumulated wisdom and the boundless insights from the cosmos gives rise to the concept of a unified universal language. This monumental endeavor transcends linguistic communication to encompass the shared essence of knowledge that permeates both the microcosm of our world and the macrocosm of the universe. Moreover, integrating terrestrial and extraterrestrial understandings embodies the pinnacle of interdisciplinary collaboration, where the boundaries of traditional academic domains blur in the pursuit of cosmic enlightenment. Such fusion of knowledge enables the reconciliation of diverse perspectives and the dissolution of conventional paradigms, fueling the evolution of human cognition. Advancements in this direction inspire the development of innovative technologies with the potential to shape the future of interstellar communication. The integration of Earth-bound knowledge with cosmic insights propels the advancement of interstellar communication methodologies, which in turn opens new frontiers for cross-planetary collaboration and exchange. The implications transcend mere scientific discourse, extending into sociocultural and philosophical realms. This harmonious integration also prompts profound ethical considerations, demanding thoughtful reflection on the implications of unifying our earthly comprehension with celestial revelations. Furthermore, as we navigate these uncharted territories of thought and discovery, it becomes imperative to uphold ethical standards that ensure equitable access to this integrated knowledge. It is essential to foster an inclusive and respectful approach that acknowledges the rich diversity of intellectual traditions and perspectives that contribute to both terrestrial and cosmic wisdom. As we strive to integrate Earth's knowledge with cosmic insights, we must remain cognizant of the ethical responsibilities associated with this endeavor, guided by principles of respect, equity, and shared benefit. Ultimately, the future prospects of integrating Earth's knowledge with cosmic insights offer an exhilarating journey into the depths of human understanding. It empowers us to chart a course towards holistic wisdom that transcends the boundaries of our world, uniting us in our collective quest for cosmic enlightenment.

Case Studies in Successful Interstellar Messages

In the exploration of interstellar communication, case studies serve as invaluable points of reference. Exemplary instances of successful messaging to extraterrestrial entities provide insight into the potential for cross-species or cross-planetary exchange. The history of such endeavors spans from pioneering efforts to contemporary initiatives. One notable case involves the transmission of the Arecibo message in 1974, a powerful binary representation conveyed via radio waves from the Arecibo Observatory in Puerto Rico toward the globular star cluster M13. The message encapsulated key elements of human civilization and biology in an attempt to communicate with potential extraterrestrial recipients. Another significant example is the Golden Record on the Voyager spacecraft, which contains multimedia representations of life and culture on Earth, intended for potential encounters beyond our solar system. Furthermore, the pulsar map etched onto the Pioneer plaques served as a symbolic representation of humanity's location within the Milky Way galaxy, aiming to narrate our cosmic existence to any sentient beings who might stumble upon it. These case studies underline the interdisciplinary nature of interstellar communication, drawing from astronomy, information theory, linguistics, and cultural anthropology. They also highlight the ethical considerations intertwined with signaling to unknown cosmic constituents, provoking profound discussions on the implications and responsibilities associated with interstellar messaging. As we scrutinize these successful endeavors, we cultivate a deeper comprehension of the intricacies involved in formulating messages that transcend astronomical distances and species boundaries. Such case studies not only propel scientific inquiry but also foster contemplation about humanity's place in the universe and the inherent desire for connection with entities beyond our earthly domain.

Conclusion: Toward a More Connected Universal Future

With the exploration and analysis of case studies in successful interstellar messages, we are poised on the threshold of a profound paradigm shift in our understanding of universal communication. The culmination of these studies has brought to light the potential for enhanced interconnectedness with extraterrestrial civilizations, as well as a deeper comprehension of the universal languages that pervade the cosmos.

As we have delved into the intricacies of various interstellar messages, it has become evident that the languages and methods of communication employed by advanced civilizations may hold invaluable knowledge and insights that could revolutionize our own technological, philosophical, and sociological frameworks. By studying and deciphering these interstellar missives, we open the door to a future where humanity is no longer bound solely to terrestrial concerns, but rather becomes an active participant in a cosmic conversation of unprecedented magnitude and significance.

The implications of embracing a more connected universal future reach far beyond the realms of scientific inquiry. This evolution in our understanding of communication extends to technological innovation, as we harness the knowledge gained from interstellar messages to develop new modes of transmitting and receiving data, encoding information, and establishing meaningful exchanges beyond our planetary confines. Moreover, the philosophical ramifications of our increased awareness of universal connectivity prompt profound contemplation of our place within the broader tapestry of existence, encouraging us to reassess our perspectives on identity, purpose, and coexistence within the cosmos.

Furthermore, the ethical considerations stemming from our exploration of universal communication demand rigorous scrutiny. As we progress toward a more connected universal future, it is imperative that we approach these endeavors with an acute awareness of our responsibilities as stewards of knowledge and ambassadors of humanity. Respect for the cultural, intellectual, and moral diversity inherent in the languages of the universe is paramount, and our interactions with potential extraterrestrial intelligences must be informed by principles of compassion, cooperation, and mutual respect.

Looking ahead, the realization of a more connected universal future invites us to consider our role in shaping the trajectory of human development and evolution. Our growing proficiency in deciphering and interpreting interstellar languages holds the promise of not only expanding the frontiers of knowledge and understanding, but also fostering profound transformations in our collective consciousness and global unity. In embracing this grand endeavor, we stand at the precipice of profound transformation, as we embark on a journey toward a future defined by interconnectedness, enlightenment, and the shared pursuit of cosmic wisdom.

Chapter 23

Interstellar Linguistics: Understanding Communication Beyond Our Solar System

Interstellar Linguistics

Interstellar linguistics is a field of interdisciplinary study that delves into the complexities of deciphering and understanding potential communication systems from extraterrestrial sources. The significance of interstellar linguistics lies in its potential to bridge the gap between humanity and possible intelligent life forms beyond our solar system. By unraveling the intricacies of non-terrestrial languages, we not only gain insights into the fundamental principles of communication but also open the door to profound discoveries about the nature of distant civilizations. This enigmatic pursuit holds the promise of expanding our comprehension of the universe, redefining our understanding of language, and broadening the scope of our existence. Understanding interstellar linguistics involves an exploration of various theoretical frameworks, technological advancements, historical context, and the profound implications it carries for humankind's future interactions with potential extraterrestrial intelligences. Additionally, this field demands rigorous analysis, critical thinking, and an open-minded approach as we endeavor to comprehend communication systems that may differ radically from any form of language encountered on Earth. The complexity and scale of interstellar linguistics present a daunting yet exhilarating challenge, as it compels us to rethink our linguistic structures, cultural presumptions, and cognitive processes. The investigation of potential linguistic communications beyond the boundaries of our planet requires a methodical, collaborative, and cross-disciplinary approach, integrating expertise from fields such as astronomy, cognitive science, anthropology, computer science, and more. As we embark on this intellectual journey, we are poised to encounter a tapestry of diverse hypotheses, speculative conjectures, and groundbreaking discoveries that have the potential to reshape our understanding of communication, intelligence, and the universe at large.

Historical Context and Early Theories

Throughout history, humanity has been captivated by the notion of extraterrestrial life and the possibility of communicating with beings from beyond our solar system. The earliest inklings of interstellar communication emerged in ancient civilizations where myths, legends, and religious texts often referenced encounters with otherworldly entities. These accounts sparked curiosity and laid the foundation for early theories on interstellar linguistics. In ancient Greece, philosophers like Epicurus proposed that there could be other worlds with inhabitants, setting the stage for further contemplation of alien communication. Moving through the annals of history, we encounter numerous thinkers and visionaries who dared to speculate about the potential existence of extraterrestrial civilizations and the methods of deciphering their languages. In the Renaissance era, the concept of plurality of worlds gained prominence, leading to discussions on the conceivable forms of Alien languages and their potential structure. This line of thought paved the way for crucial developments in the field of astronomy, as scholars made strides in understanding the vastness of the universe and the likelihood of intelligent life beyond Earth. As scientific knowledge expanded, so did the theories surrounding communication with extraterrestrial life. The advent of radio technology in the 19th century ignited new hope for intercepting signals from outer space, spurring interest in the search for cosmic messages. Early pioneers like Nikola Tesla and Guglielmo Marconi entertained ideas of receiving interstellar transmissions, planting the seeds for future pursuits in interstellar communication. The dawn of the 20th century marked a turning point in the study of interstellar linguistics, with scientific advancements and interdisciplinary collaborations propelling the pursuit of understanding non-terrestrial languages. Visionaries such as Carl Sagan and Frank Drake laid the groundwork for SETI (Search for Extraterrestrial Intelligence) and formulated the renowned Drake Equation to estimate the number of communicative civilizations in our galaxy. Historical context and the evolution of early theories in interstellar linguistics provide valuable insights into the human fascination with communicating across cosmic distances and set the stage for the modern exploration of understanding communication beyond our solar system.

Decoding the Cosmic Lexicon: Theories and Methods

As we delve into the intricate realm of interstellar linguistics, the pursuit of deciphering the cosmic lexicon demands a comprehensive understanding of the theories and methods that underpin this profound endeavor. The search for extraterrestrial intelligence is rooted in centuries of human fascination with the cosmos, and theories regarding potential forms of cosmic communication have evolved alongside advancements in scientific inquiry and technological innovation. Various theoretical frameworks have been proposed to model potential extraterrestrial languages, ranging from mathematical concepts to the study of universal principles in communication. Additionally, the field of astrobiology provides essential insights into the potential biochemistry and cognitive capabilities of hypothetical alien civilizations, informing the design of methods aimed at decoding their communications. Central to this pursuit is the development of interdisciplinary methodologies that integrate insights from linguistics, anthropology, information theory, and astronomy. These methodologies encompass both empirical and theoretical approaches, leveraging advances in fields such as machine learning, computational linguistics, and signal processing to analyze complex data sets obtained from astronomical observations. Furthermore, the advent of sophisticated telescopic arrays and deep-space probes has facilitated the collection of vast amounts of data, enabling researchers to explore nuanced patterns and anomalies that may signify deliberate attempts at communication. Moreover, the utilization of advanced algorithms and artificial intelligence enhances our capacity to discern potentially meaningful signals from the natural background noise inherent in the universe. Integrating these technological resources with robust linguistic analysis allows for the systematic exploration of fundamental parameters that would distinguish intelligent communication from random cosmic phenomena. In the pursuit of decoding the cosmic lexicon, it is imperative to remain cognizant of the complexities and uncertainties involved in inferring meaning from interstellar transmissions. Given the profound implications of successful communication with extraterrestrial civilizations, ethical considerations and responsible protocols for interaction must be integrated into the overarching framework of interstellar linguistics. As we survey the horizon of possibilities, addressing the challenges and bearing in mind the ethical dimensions of this quest will guide us toward a more comprehensive understanding of the cosmic lexicon and its potential significance for humanity.

Technological Advances in Extraterrestrial Communication

Advances in technology have significantly enhanced humanity's capability to engage in the search for extraterrestrial intelligence and communicate with potential alien civilizations. The evolution of communication technology, including radio telescopes and deep space probes, has revolutionized our capacity to reach out to the cosmos and listen for signals from distant stars. State-of-the-art instruments such as the Very Large Array (VLA) and the Arecibo Observatory enable scientists to capture and analyze radio emissions from celestial bodies, expanding the boundaries of our understanding of the universe. Additionally, advancements in signal processing algorithms and computational power have empowered researchers to sift through vast amounts of data in real time, seeking patterns and anomalies that may indicate intentional interstellar transmissions.

Furthermore, the development of laser-based communication technologies has opened up new channels for potential contact with extraterrestrial civilizations. Projecting powerful laser beams into space has the potential to create detectable signals that could be intercepted by advanced alien races. Through initiatives such as Breakthrough Starshot, which aims to develop ultra-fast light-driven nanocraft, we are on the verge of a new era in interstellar communication where the speed and reach of information exchange could be unparalleled.

Moreover, quantum communication research holds promise for establishing secure and efficient means of communicating with hypothetical extraterrestrial counterparts. Harnessing the principles of quantum entanglement and cryptography, scientists are exploring the potential for developing quantum communication protocols that could resist interception and decryption by any eavesdropping entities, terrestrial or otherwise. This presents a transformative opportunity for establishing trustful and confidential communication channels across cosmic distances, laying the groundwork for substantial progress in extraterrestrial contact scenarios.

These technological strides not only empower our search for extraterrestrial intelligence but also elevate our ability to conduct long-term dialogue with potential alien civilizations. As human civilization continues to advance technologically, these developments serve as indispensable components of our ongoing quest to unravel the mysteries of the cosmos and establish meaningful connections with other intelligent beings in the universe.

The Search for Extraterrestrial Intelligence (SETI): Efforts and Findings

As humanity continues to explore the cosmos, one of the most captivating pursuits has been the search for extraterrestrial intelligent life. The Search for Extraterrestrial Intelligence (SETI) represents a collaborative effort by scientists, researchers, and enthusiasts to detect potential signals or evidence of advanced civilizations beyond our world. SETI initiatives have been pivotal in shaping our understanding of the universe and have led to numerous fascinating findings. One of the central efforts of SETI has involved scanning the vast expanse of space for anomalous electromagnetic signals that could potentially be indicative of non-terrestrial origin. Through meticulously designed telescopic arrays and sensitive monitoring equipment, scientists have dedicated countless hours to sifting through the cosmic radio waves in search of a signal that defies natural explanations. While no conclusive evidence of extraterrestrial intelligence has been documented, SETI endeavors have unearthed intriguing anomalies and transient signals that continue to fuel scientific discourse. Exemplifying the dedication to this field, the SETI community consistently collaborates on global scales, utilizing cutting-edge technology and innovative methodologies to expand the frontiers of cosmic exploration. Despite the absence of definitively artificial signals, the perseverance and unwavering commitment of the SETI community underscore the enduring quest to unravel the mysteries of the cosmos. Furthermore, the proliferation of innovative algorithms and advanced computing systems has empowered researchers to process unprecedented volumes of data, enabling more comprehensiv e and efficient analyses in the pursuit of detecting potential extraterrestrial transmissions. Beyond conventional radio wave detection, SETI has diversified its approach to include searches for optical signals, spectral irregularities, and even interstellar artifacts, broadening the scope of exploration to encompass a myriad of communication modalities. While each endeavor may not yield immediate results, the cumulative progress serves as a testament to the collective determination to fathom the enigma of extraterrestrial intelligence. The unprecedented collaboration within the SETI community not only advances scientific knowledge but also inspires a sense of wonder and contemplation that transcends disciplinary boundaries. Through this unyielding pursuit, humankind embraces the profound curiosity to comprehend our place in the cosmos and seeks to establish meaningful connections with potential sentient beings beyond the confines of Earth.

Linguistic Patterns Observed in Known Signals

Space exploration and the search for extraterrestrial intelligence have long captured the imagination of scientists and the public alike. Among the myriad challenges that scientists face in decoding potential messages from distant civilizations is the identification and interpretation of linguistic patterns observed within known signals. While these patterns may not conform to terrestrial languages, there is an underlying hope that they may provide crucial insights into the nature of communication beyond our own planet.

In analyzing known signals received through initiatives such as the SETI program, researchers have striven to identify recurring motifs, cadences, and syntactic structures that could signify intentional communication. Despite the inherent limitations of such exercises, investigators have made substantial progress in discerning potential linguistic elements within the detected transmissions. These linguistic patterns often manifest as repetitive sequences, rhythmic variations, or distinct irregularities within the electromagnetic spectra.

Furthermore, the field of computational linguistics has played a pivotal role in identifying and categorizing these observed linguistic patterns. By leveraging advanced algorithms and machine learning techniques, scientists have been able to analyze the complex data generated by incoming signals and discern potential linguistic components. This alignment with cutting-edge technology underscores the interdisciplinary nature of this endeavor, bringing together scholars from diverse fields including astrophysics, linguistics, and computer science.

At the crux of this endeavor lies the daunting task of differentiating between naturally occurring cosmic phenomena and deliberate attempts at interstellar communication. Deciphering the intended meaning behind these observed linguistic patterns remains one of the most pressing challenges of modern science. Furthermore, the absence of a common reference point or shared cultural context with putative extraterrestrial communicators adds layers of complexity to the interpretative process. As researchers continue to sift through the troves of incoming data, their efforts are propelled by the prospect of establishing meaningful dialogues with civilizations beyond our solar system.

Ultimately, the recognition and understanding of linguistic patterns within known signals stand as pivotal milestones in humanity's quest to unravel the mysteries of the cosmos. While the road ahead is fraught with uncertainties and complexities, it is through methodical analysis and collective collaboration that we inch ever closer to comprehending potential forms of non-terrestrial intelligence.

Challenges in Understanding Non-Terrestrial Languages

As humanity delves deeper into the realm of interstellar communication, the formidable challenges posed by understanding non-terrestrial languages come to the forefront. One of the primary hurdles is the inherent limitation of human linguistic frameworks when encountering extraterrestrial modes of communication that may be fundamentally distinct from our own. This necessitates a radical shift in perspective, requiring scholars and researchers to break free from the constraints imposed by Earth-based languages and embrace a wholly novel approach to comprehension.

Furthermore, the absence of a shared cultural context or referential framework with extraterrestrial civilizations introduces an additional layer of complexity. Without common touchpoints or cultural markers, deciphering the nuances of non-terrestrial languages becomes an intricate puzzle, perpetually shrouded in ambiguity and uncertainty. The lack of context poses a significant obstacle, as meaning in language is deeply intertwined with the cultural and social milieus of its speakers. Consequently, efforts to unravel the intricacies of extraterrestrial linguistic structures are impeded by this profound disconnect.

In addition, the very concept of language, as conceived on Earth, may not align with the diverse modes of communication prevalent in advanced alien societies. Conventional assumptions about grammar, syntax, and semantics may prove inadequate in the face of non-terrestrial linguistic systems that operate on principles beyond the scope of terrestrial paradigms. This necessitates a comprehensive reevaluation of the fundamental tenets of linguistics and a willingness to entertain the possibility of entirely unprecedented communicative modalities.

Moreover, the inherent variability of non-terrestrial communication systems presents a formidable challenge. Alien languages may exhibit diverse forms, incorporating visual, auditory, olfactory, or even tactile components that expand the boundaries of traditional linguistic analysis. This multisensory dimension adds layers of complexity, necessitating interdisciplinary collaboration across fields such as cognitive science, neurobiology, and xenolinguistics to construct a comprehensive framework for interpretation.

Lastly, ethical considerations emerge as a critical concern in the pursuit of comprehending non-terrestrial languages. The potential impact of linguistic interactions on extraterrestrial societies demands a responsible and introspective approach, with careful deliberation on the consequences of our endeavors. Unintentional misinterpretations or miscommunications could carry profound repercussions, underscoring the need for sensitivity, humility, and ethical discernment in the quest to unlock the enigmatic languages of the cosmos.

Navigating these profound and multifaceted challenges demands an unparalleled dedication to scientific rigor, openness to paradigm-shifting insights, and a deep appreciation for the profound complexities inherent in the cosmic tapestry of languages. It is through acknowledging and confronting these challenges that humanity can aspire to bridge the interstellar communicative chasm and engage in meaningful dialogue with civilizations beyond the confines of our celestial shores.

Simulating Alien Languages: Approaches and Experiments

As humanity advances in its exploration of space, the question of interstellar communication becomes increasingly relevant. The challenge of understanding non-terrestrial languages has led to the development of innovative approaches and experiments aimed at simulating and deciphering potential alien languages. The field of xenolinguistics, the study of hypothetical alien languages, has emerged as a fascinating interdisciplinary pursuit that combines linguistics, cognitive science, anthropology, and astrophysics. One approach to simulating alien languages involves leveraging computational linguistics to create artificial agents capable of generating and interpreting linguistic signals based on imagined extraterrestrial syntax and semantics. By modeling the cognitive processes involved in language acquisition and understanding, researchers hope to simulate plausible alien language structures and meanings. Another avenue of inquiry lies in conducting thought experiments and simulations based on speculative principles of communication that may apply to extraterrestrial civilizations. These simulations involve hypothesizing about the biological, environmental, and technological factors that could shape an alien species' methods of communication. Furthermore, experimental simulations utilizing advanced neural networks and machine learning algorithms are being devised to analyze and interpret potential extraterrestrial signals that may exhibit patterns indicative of intelligent origin. Such experiments aim to expose these systems to diverse linguistic stimuli and observe how they recognize and process information encoded in non-human modes of communication. Despite the inherent challenges and uncertainties associated with simulating alien languages, these endeavors serve as critical exercises in expanding our understanding of the possible forms of interstellar communication. Additionally, experimental approaches include studying the communication systems of terrestrial non-human species, such as dolphins, whales, and certain primate species, to gain insights into alternative forms of linguistic and communicative expression. These comparative studies provide valuable perspectives that inform our theoretical frameworks for simulating and comprehending potential extraterrestrial languages. By elucidating the principles underlying diverse modes of communication on Earth, we cultivate a richer conceptual toolkit for approaching the complexities of simulating and interpreting alien languages in the cosmos. Moreover, interdisciplinary collaboration and knowledge-sharing between linguists, astronomers, neuroscientists, computer scientists, and ethicists play a crucial role in shaping the methodologies and ethical considerations guiding the simulation and analysis of potential alien languages. As the quest for interstellar communication progresses, the integration of diverse expertise and the ethical navigation of this uncharted territory will be pivotal in ensuring the responsible and meaningful pursuit of understanding and potentially interacting with extraterrestrial intelligences.

Ethical Considerations in Interstellar Communication

As humanity delves into the realm of interstellar communication, it is crucial to confront the ethical considerations inherent in attempting to establish contact with extraterrestrial civilizations. The implications of ini-

tiating communication with alien species are profound and far-reaching, transcending scientific and technological domains to encompass moral, philosophical, and existential dimensions. Ethicists, scientists, policymakers, and the broader public must collectively grapple with a myriad of complex issues as we venture into this uncharted territory. At the forefront of ethical deliberations is the imperative to approach potential alien encounters with respect, humility, and caution. Respect for the autonomy and sovereignty of extraterrestrial beings is paramount, necessitating careful contemplation of the potential consequences of our actions. Moreover, considerations of the cultural, social, and psychological impact on both human and alien societies must be thoroughly examined. There exists the possibility that unintentional harm or misunderstanding may arise from our interactions with extraterrestrial entities, highlighting the need for stringent ethical guidelines and protocols. Furthermore, questions pertaining to the dissemination of information about Earth and humanity, and the potential vulnerabilities and risks associated with such disclosures, demand rigorous ethical examination. The ethical dilemmas surrounding the potential discovery of alien knowledge, technology, and biological information also warrant thoughtful reflection. Preservation of biodiversity and the prevention of inadvertent contamination arising from interstellar communication efforts serve as additional ethical imperatives. Engaging in open and transparent dialogue within the global community is essential in navigating these ethical complexities. Our collective responsibility to act as stewards of the Earth and representatives of humanity demands that we approach interstellar communication with wisdom, prudence, and a deep commitment to ethical considerations. As we continue to pursue avenues for interstellar communication, it is incumbent upon us to uphold the highest ethical standards, fostering a global discourse that honors the values, aspirations, and welfare of terrestrial and extraterrestrial civilizations alik
e.

Future Prospects and Next Steps in Interstellar Linguistics

The exploration of interstellar linguistics presents a compelling opportunity for humanity to expand its understanding of the universe and our place within it. As technology continues to advance, unlocking the complexities of non-terrestrial languages becomes increasingly feasible. Looking ahead, there are several key factors that will shape the future prospects and next steps in interstellar linguistics.

Firstly, advancements in artificial intelligence and machine learning hold immense promise for deciphering extraterrestrial communication. By leveraging sophisticated algorithms and data processing capabilities, researchers can analyze vast sets of cosmic signals and patterns with greater efficiency and accuracy. Collaborations between computational linguists, astrophysicists, and cognitive scientists are poised to yield groundbreaking insights into the structure and semantics of alien languages.

Moreover, international collaboration and interdisciplinary research initiatives will be pivotal in driving progress in the field of interstellar linguistics. Establishing global networks of linguists, astronomers, and technologists will foster a collective approach to addressing the multifaceted challenges posed by non-terrestrial communication. Open data sharing and cooperative projects will accelerate the pace of discovery and enhance our ability to decipher enigmatic celestial messages.

Another crucial facet of the future of interstellar linguistics lies in the ethical considerations surrounding potential contact with extraterrestrial civilizations. As humanity ventures closer to the prospect of communicating with intelligent beings from other worlds, ethical frameworks for interstellar diplomacy and respectful engagement must be developed. Scholars and policymakers will need to collaborate in establishing guidelines for responsible interaction and information exchange, ensuring that any communication attempts are executed with prudence and sensitivity.

Looking forward, the prospect of harnessing quantum communication technologies for interstellar messaging represents a transformative frontier in the field. Quantum entanglement and encryption techniques offer the potential for instantaneous, secure communication across vast cosmic distances, revolutionizing our capacity to engage in meaningful dialogue with extraterrestrial entities. This paradigm shift in communication methods holds the promise of overcoming the limitations imposed by traditional electromagnetic signals, opening new avenues for interstellar linguistic exploration.

In conclusion, the future of interstellar linguistics is imbued with profound scientific, philosophical, and existential implications. As humanity stands on the cusp of potentially unraveling the languages of distant civilizations, the responsible pursuit of this knowledge will define our shared legacy. By embracing technological innovation, international cooperation, ethical deliberation, and pioneering communication methods, we are poised to usher in a new era of interstellar understanding that transcends terrestrial boundaries and expands the frontiers of human cognition.

Chapter 24

Epilogue: Reflections on Interstellar Linguistics

Throughout our exploration of the diverse languages and signals of the universe, we have encountered a remarkable continuity that underscores the interconnectedness of cosmic communication. The rich tapestry of Earth's linguistic and communicative systems has provided invaluable insights into the potential for synthesizing our understanding with the enigmatic signals emanating from the cosmos.

The parallels between the intricate codes of nature, as elucidated in earlier chapters, and the elusive transmissions from distant celestial bodies are not mere coincidences but rather reflections of a universal framework of communication. Just as we unraveled the nuanced dialects of animals and deciphered the ancient narratives embedded within geological formations, we now stand poised at the threshold of comprehending the profound messages conveyed by the cosmos.

The interplay between these different layers of communication—be it the subtle cues of plant life or the monumental symphonies of celestial phenomena—reveals a profound unity in the language of the universe. Our multidisciplinary journey has illuminated the potential for an integrated approach to interstellar linguistics, wherein the patterns of Earth's countless dialects provide a foundation for interpreting the enigmatic discourse of the stars.

Moreover, this synthesis of Earth's languages with cosmic signals holds the promise of transcending the limitations of terrestrial boundaries, offering a glimpse into the broader understanding of universal language. By weaving together the threads of knowledge gleaned from terrestrial ecosystems and applying them to the analysis of extraterrestrial communications, we seek to align our collective consciousness with the unspoken dialogue that permeates the cosmos.

As we reflect on the convergence of these diverse sources of communication, we are compelled to acknowledge the indispensable role of interdisciplinary collaboration in unraveling the fabric of interstellar linguistics. By uniting the expertise of linguists, astronomers, biologists, and geologists, we embark on a collective endeavor to discern the underlying grammar of the universe and compose a lexicon that transcends the confines of individual disciplines.

Intertwined with these reflections is the recognition of a fundamental shift in our understanding of language itself, as it ceases to be confined to the familiar cadences of human speech and extends into the boundless expanse of the cosmos. This awakening inspires us to contemplate the intricate web of connections that binds us to the cosmos, fostering an awareness of our place in a grand narrative of cosmic communication that transcends the confines of time and space.

In illuminating the reflections on interstellar linguistics, we are poised on the cusp of a profound realization—a realization that extends far beyond the empirical foundations of our exploration and delves into the very essence of our existence as sentient beings engaged in a timeless conversation with the universe.

Synthesizing Earth's Languages with Cosmic Signals

Humanity has long grappled with the idea of communicating with extraterrestrial beings and interpreting cosmic signals. While our understanding of Earth's languages and natural communication systems has progressed significantly, the prospect of synthesizing these intricacies with cosmic signals presents a unique set of challenges and opportunities. This endeavor necessitates a holistic approach that draws upon diverse fields such as linguistics, astronomy, information theory, and cognitive science.

At the intersection of earthly and cosmic languages lies the intricate task of identifying commonalities and establishing frameworks for mutual comprehension. Given the diversity of terrestrial languages, from spoken and written forms to visual and non-verbal cues, discerning universal patterns that transcend cultural and species boundaries is paramount. Moreover, decoding cosmic signals requires an appreciation of potential modalities beyond traditional human-centric communication methods, encompassing variations in signal frequency, amplitude, temporal patterns, and spatial encoding.

Furthermore, the synthesis of Earth's languages with cosmic signals demands a multidisciplinary exploration of semiotics, the study of signs and symbols. By delving into the fundamental principles governing the creation and interpretation of meaning, researchers seek to develop a framework for deciphering the language of the cosmos. This pursuit involves investigating the semiotic structures and semantic content embedded within terrestrial communications and mapping these onto hypothetical extraterrestrial transmissions.

In this context, artificial intelligence and machine learning technologies have emerged as indispensable tools in the quest to synthesize Earth's languages with cosmic signals. These advanced computational approaches play a pivotal role in pattern recognition, classification, and semantic analysis, offering insight into the underlying structures of linguistic and non-linguistic expressions. Through the application of algorithms capable of identifying recurring motifs and deciphering complex syntax, the potential exists to discern commonalities between Earth-bound languages and cryptic cosmic transmissions.

As we navigate the terrain of synthesizing Earth's languages with cosmic signals, it becomes evident that this undertaking transcends the realm of pure scientific inquiry. It prompts contemplation of our place in the universe, our aspirations for cosmic communion, and the ethical implications associated with potential interstellar dialogue. The assimilation of these multifaceted considerations into interdisciplinary dialogues represents a critical step forward in humanity's collective efforts to engage in meaningful exchanges with the cosmos.

The Role of Human Perception in Universal Communication

Human perception plays a crucial role in the quest to decipher and comprehend universal communication. As we strive to understand cosmic languages, our very ability to perceive and interpret signals from extraterrestrial sources becomes an essential component of achieving meaningful interaction with potential intelligences beyond our world. The limitations and biases inherent in human perception must be carefully considered as we endeavor to bridge the gap between Earth's languages and those of the cosmos. Cognitive anthropologists, linguists, and experts in psychology and neurobiology are all essential in this pursuit. They provide insights into the intricacies of human sensory systems, information processing, and social cognition, shedding light on how our own perceptions may shape our interpretations of interstellar communications. Furthermore, interdisciplinary collaborations with researchers specializing in artificial intelligence, machine learning, and data analytics offer promising avenues for refining our understanding of universal languages. By leveraging cutting-edge technologies, such as advanced signal processing algorithms and pattern recognition software, we can enhance our capacity to detect, decode, and comprehend enigmatic cosmic transmissions. Moreover, these tools enable us to address the challenges posed by potential variations in communication modalities that may differ fundamentally from those encountered within the terrestrial sphere. In tandem with technological advancements, educating and engaging the public on the intricacies of universal communication is imperative. Cultivating an appreciation for the nuances and complexities of interpreting cosmic messages fosters a more informed and receptive global community, equipped to contribute meaningfully to the pursuit of unraveling the multifaceted tapestry of universal dialogue. This educational outreach also serves to cultivate cultural preparedness for potential encounters with extraterrestrial civilizations, offering a platform for discourse on the ethical, social, and existential implications of such profound discoveries. Ultimately, the role of human perception in universal communication encompasses not only the scientific exploration of cognitive processes and technological innovations but also the societal and philosophical dimensions that accompany our collective journey towards embracing our place in the wider cosmic convers ation.

Advancements in Technology and their Impact on Cosmic Understanding

The quest to comprehend the cosmic landscape has been significantly influenced by technological advancements. From the invention of telescopes to cutting-edge instruments like the Hubble Space Telescope, technology has expanded our ability to perceive and interpret the universe. Imaging technologies have allowed us to capture stunning visuals of celestial bodies and phenomena, shedding light on the mysteries of the cosmos. Meanwhile, spectroscopy techniques have enabled us to analyze the chemical composition of distant stars and galaxies. These technologies have fundamentally altered our comprehension of the universe, allowing us to delve into the minute details of its vast expanse. Furthermore, the development of space probes and rovers has provided firsthand data from other planets and moons, offering invaluable insights into the geology, atmospheric conditions, and potential for extraterrestrial life. Advancements in radio astronomy have further revolutionized our understanding by detecting cosmic signals and studying the cosmic microwave background radiation, which provides essential clues about the early universe. The application of artificial intelligence and machine learning algorithms in processing astronomical data has expedited analysis and pattern recognition, facilitating discoveries that were previously unimaginable. Moreover, the utilization of high-performance computing has supported complex simulations and modeling, allowing researchers to test hypotheses and simulate cosmic phenomena under various conditions. As we embrace the era of big data, interdisciplinary collaborations between astronomers, astrophysicists, engineers, and data scientists are widening our horizons and opening new frontiers in cosmic exploration. Consequently, these technological leaps are propelling humanity towards a more profound understanding of the universe, making it an exciting time for cosmological research and discov ery.

Philosophical Implications of a Unified Language Theory

The quest to understand and interpret universal languages extends beyond the realm of science; it delves into the heart of philosophy, sparking profound contemplation on humanity's place in the cosmos. A unified language theory, if actualized, would revolutionize our perception of existence, challenging conventional boundaries of communication and knowledge. This paradigm shift invites exploration into the philosophical implications of such a transformative concept. At the core of this discourse lies the age-old query regarding human uniqueness. With the potential revelation of a universal language, the definition of what makes us distinct as a species may be redefined. The unification of linguistic principles across cosmic boundaries prompts us to reconsider our perceived isolation in the universe, urging us to confront the reality of our interconnectedness with other intelligences. The ethical dimensions of this proposition are equally profound. As we strive to decipher and engage in interstellar communication, we are compelled to appraise our moral responsibility as custodians of Earthly wisdom. Such endeavors provoke questions about the dissemination of knowledge, respect for cultural diversity, and the preservation of indigenous languages in light of potential galactic interactions. Moreover, the prospect of universal communication challenges our conception of truth and reality. How will the unification of languages impact our understanding of fundamental concepts such as morality, consciousness, and the nature of existence? Could convergence in linguistic interpretation lead to a collective expansion of consciousness that transcends individual or planetary perspectives? These contemplations necessitate introspection into the very essence of the human experience and its relationship with the wider cosmic tapestry. Furthermore, a unified language theory beckons a reevaluation of our cosmic significance. It propels us beyond terrestrial preoccupations, fostering a deeper comprehension of our place in the grand narrative of the universe. As we bridge the gap between earthly dialects and potential extraterrestrial languages, we are compelled to reassert our role as stewards of knowledge and wisdom in the greater cosmic community. Consequently, this pursuit engenders a poignant reflection on the purpose and destiny of humanity within the vast expanse of the cosmos. Through a philosophical lens, the notion of a unified language theory provokes a renaissance of thought, encouraging multidisciplinary dialogue and inspiring a collective contemplation of our cosmic identity.

Future Research Directions in Universal Linguistics

As we stand on the precipice of unprecedented advancement in our understanding of universal linguistics, it becomes imperative to contemplate the potential avenues for future research. The exploration of extraterrestrial communication holds profound implications not only for scientific discovery but also for the fundamental comprehension of humanity's place in the cosmos. One compelling direction for future research involves the development of advanced algorithms and machine learning systems to decipher and interpret complex interstellar signals. These sophisticated tools have the potential to unlock the intricacies of alien languages, paving the way for meaningful dialogue with civilizations beyond our planet. Furthermore, interdisciplinary collaborations are essential to propel the field forward. By fostering partnerships between linguists, astronomers, anthropologists, and technologists, we can harness diverse expertise to tackle the multifaceted challenges posed by the study of universal linguistic codes. This integrative approach presents an opportunity to examine extraterrestrial communication through a comprehensive lens, enabling us to tease apart the nuances of cosmic languages. Additionally, the establishment of a global network for the collection and analysis of interstellar data is a crucial aspect of future research in universal linguistics. By uniting observational efforts across continents and leveraging state-of-the-art telescopic arrays, researchers can construct a more extensive and detailed database of extraterrestrial signals. This collaborative initiative would not only facilitate the rapid identification of potential communication from distant galaxies but also allow for longitudinal studies of cosmic language patterns. Moreover, ethical considerations form an indispensable component of future research directions in universal linguistics. As we delve deeper into the realm of extraterrestrial communication, it is vital to contemplate the ethical dimensions of our pursuits. Conscientious frameworks for the responsible interpretation and dissemination of alien messages must be developed to ensure that our interactions with possible extraterrestrial intelligences adhere to principles of respect and mutual understanding. Finally, future research in universal linguistics should strive to preserve and celebrate Earth's rich linguistic heritage while navigating the intricacies of cross-species and interplanetary communication. By drawing inspiration from the myriad terrestrial languages and dialects that have evolved on our planet, researchers can cultivate a deeper appreciation for the diversity of linguistic expression, thereby enriching our engagement with potential extraterrestrial languages. In sum, the pursuit of future research directions in universal linguistics demands a concerted effort towards technological innovation, interdisciplinary collaboration, ethical contemplation, and cultural reverence. It is through such endeavors that we may aspire to unravel the enigmatic tapestry of cosmic communication and embrace the profound interconnectedness of linguistic dialogues within and beyond our celestial boundaries.

Collaborative Efforts Across Disciplines and Civilizations

In the pursuit of deciphering the universal language, collaborative efforts play a pivotal role in leveraging diverse expertise and resources. Interdisciplinary cooperation allows for a holistic approach that transcends the confines of individual fields, fostering an environment where specialists from various disciplines can synergize their strengths and address the complexities of universal linguistics. The fusion of astrophysics, linguistics, anthropology, technology, and ethics is essential in unraveling the enigma of extraterrestrial communication. By pooling together knowledge, methodologies, and perspectives, researchers have the opportunity to enhance their understanding and interpretation of cosmic signals. Moreover, collaboration between different civilizations is imperative in broadening the scope of inquiry and promoting cultural exchange. International consortia dedicated to the search for extraterrestrial intelligence facilitate a global exchange of ideas and strategies, enriched by the diversity of cultural and scientific backgrounds. Furthermore, collaborations across civilizations not only provide a platform for shared learning and innovation but also foster diplomatic relations, promoting peaceful interactions on a global scale. As humanity embarks on the quest to decode the universal language, it becomes evident that such an endeavor demands concerted efforts and inclusive participation from disparate disciplines and civilizations. This cooperative ethos will be instrumental in establishing a comprehensive framework for universal linguistic research, transcending borders and boundaries to embrace the collective knowledge and wisdom of humankind in our exploration of the cosmos.

Ethical Considerations in the Search for Extraterrestrial Intelligence

As humanity continues its quest to unlock the secrets of the universe, the search for extraterrestrial intelligence raises complex ethical considerations. The pursuit of contact with intelligent life beyond our planet presents a myriad of philosophical and moral dilemmas that demand careful deliberation. One of the foremost ethical concerns revolves around the potential impact on indigenous extraterrestrial civilizations. The inadvertent disruption or interference with alien cultures, belief systems, and socio-political structures must be approached with the utmost caution and sensitivity. Furthermore, the prospect of influencing or being influenced by advanced extraterrestrial societies prompts examination of our own societal values, governance structures, and technological advancements. Another critical ethical facet involves the responsible utilization of resources for exploring extraterrestrial intelligence. Careful allocation of scientific funding, preservation of environmental integrity, and consideration for competing human needs are paramount in the ethical pursuit of such endeavors. Additionally, the establishment of guidelines for diplomatic, scientific, and commercial interactions with potential extraterrestrial entities is essential for maintaining ethical conduct in the face of unprecedented discovery. The ethical implications of any potential communication with extraterrestrial intelligence also necessitate reflection on the impact it may have on terrestrial religions, worldviews, and socioeconomic stability. Contemplation of the emotional, psychological, and existential ramifications on the global populace is fundamental when preparing for potential paradigm shifts resulting from interstellar communication. Moreover, the ethical responsibility of transparency and truthfulness in reporting findings related to extraterrestrial intelligence cannot be overstated. Ensuring open and honest dissemination of information, while mitigating unnecessary panic or social upheaval, is imperative for sustaining public trust and societal coherence. Finally, the broaching of ethical issues in the search for extraterrestrial intelligence calls for a profound reconsideration of our place in the universe. This necessitates an introspective evaluation of our species' interconnectedness with Earth's biosphere, as well as the broader galactic community. Addressing these ethical considerations demands global collaboration among scientists, ethicists, policymakers, and representatives of diverse cultural backgrounds to create a framework that prioritizes respect for all forms of sentient life and the preservation of cosmic diversity.

Preserving Earth's linguistic heritage in the Galactic Context

Humanity has long been captivated by the idea of communicating with extraterrestrial beings and unraveling the mysteries of the cosmos. As we venture deeper into the realms of interstellar communication and exploration, an important ethical consideration arises - the preservation of Earth's rich linguistic heritage in the context of potential galactic encounters. This preservation is not only crucial for understanding our own history and identity but also for ensuring that our diverse languages and cultures are respected and preserved in any future interactions with extraterrestrial civilizations. The need to safeguard our linguistic heritage in the galactic context presents a unique set of challenges and opportunities. One of the fundamental challenges is the vast diversity of languages and dialects spoken on Earth. The task of documenting, cataloging, and archiving these languages, some of which are at risk of extinction, becomes increasingly vital as we contemplate the possibility of encountering alien forms of communication. In the event of contact with extraterrestrial intelligence, our linguistic diversity could serve as a valuable resource for comparison and analysis, potentially providing insights into universal patterns of communication and cognition. Moreover, preserving Earth's linguistic heritage extends beyond the purely scientific or practical realm; it holds profound cultural and philosophical significance. Our languages encapsulate centuries of human experience, wisdom, and creativity. They reflect the unique worldviews, values, and traditions of diverse communities across the globe. Preserving this mosaic of languages and ensuring their representation in potential interstellar discourse is a testament to the richness of human expression and a celebration of our shared planetary legacy. Embracing the diversity of Earth's languages within the galactic context also underscores the importance of cultural sensitivity, empathy, and inclusion in our interactions with hypothetical extraterrestrial civilizations. Just as we acknowledge the value of linguistic diversity on our own planet, we must approach the prospect of cosmic communication with respect for the cultural and linguistic diversity that may exist beyond Earth. This inclusive mindset not only fosters goodwill and mutual understanding but also serves as a reflection of our moral and ethical responsibility as terrestrial ambassadors to the stars. While the pursuit of preserving Earth's linguistic heritage in the galactic context is complex, it offers unparalleled opportunities for interdisciplinary collaboration and global cooperation. From linguists and anthropologists to astronomers and philosophers, diverse fields can converge to address the multifaceted dimensions of this endeavor. By fostering international partnerships and leveraging technological innovations, we can create comprehensive initiatives aimed at safeguarding and promoting linguistic diversity both on Earth and in the cosmic arena. Furthermore, the effort to preserve Earth's linguistic heritage in the galactic context calls for an open dialogue on the ethics of interstellar communication and cultural exchange. The questions that arise compel us to explore the depths of our own identities and values while simultaneously contemplating our place in the wider tapestry of the cosmos. As we embark on this journey of preservation, we embrace the responsibility of honoring the voices of our past and present, cognizant of the potential impact on the unfolding narrative of our collective future.

Conclusion: Toward a Greater Understanding of Our Place in the Cosmos

The exploration of linguistic heritage within the context of the cosmos imparts profound significance to the narrative of human endeavor. As we strive to decipher the universal language and navigate the intricate nuances of communication beyond our world, it becomes imperative to reflect on the implications of this pursuit. Through the preservation of Earth's linguistic legacy and the analysis of cosmic symbolism, humanity can endeavor to achieve a greater understanding of its place in the vast expanse of the universe. This monumental journey requires a harmonious blend of scientific inquiry, philosophical contemplation, and ethical conscientiousness. It beckons us to transcend cultural, national, and planetary boundaries to seek unity amidst the diversity of languages spoken by the celestial chorus. The preservation of our linguistic heritage in the galactic tapestry serves as a bridge between our terrestrial origins and the uncharted frontiers of interstellar communication. By safeguarding the multifaceted dialects that have emerged from the cradle of Earth, we honor the diverse narratives written by countless civilizations across epochs. Furthermore, as we contemplate the convergence of languages from distant worlds, we are compelled to ponder the ethical responsibilities inherent in our pursuit of extraterrestrial intelligence. Respect, empathy, and humility must guide our actions as we navigate the delicate balance between curiosity and respect for otherworldly cultures. The implications of deciphering the universal language permeate the realm of philosophy, provoking existential inquiries about our place in the grand design of the cosmos. They invite introspection on the interconnectedness of life forms across the stars and the moral imperatives associated with our interactions within the cosmic community. Intricately intertwined with these reflections is the pivotal role of technological advancements in amplifying our understanding of cosmic communications. From radio telescopes to deep space probes, human innovation has propelled us toward the threshold of interstellar linguistic revelations. The development of advanced algorithms and machine learning models has equipped us with the tools to decode the enigmatic messages embedded in cosmic transmissions. Notably, as we continue to refine these technological marvels, we must remain vigilant in aligning progress with ethical considerations, nurturing a symbiosis of knowledge and empathy as we probe the frontiers of the unknown. Ultimately, the journey toward a greater understanding of our place in the cosmos impels humanity to embark on a collaborative odyssey that transcends temporal, spatial, and ideological barriers. This odyssey unfolds at the intersection of empirical research, philosophical introspection, ethical discernment, and technological innovation. It calls upon us to embrace the enigma of the universe with prudence and reverence, weaving together the threads of linguistic heritage into the ever-evolving narrative of our cosmic voyage.

www.ingramcontent.com/pod-product-compliance
Ingram Content Group UK Ltd.
Pitfield, Milton Keynes, MK11 3LW, UK
UKHW021712190726
13853UKWH00001B/497

9 798224 895731